Section Five — The Atom and Radioactivity

Section Six — Generating Electricity

Section Seven — Medical Applications of Physics

Section Eight — Mechanics and Gases

Section Nine — Waves and Lenses

Section Ten — The Earth and the Universe

Section Eleven — Electricity for Gadgets

Published by CGP

From original material by Richard Parsons.

Editors:
Helena Hayes, Felicity Inkpen, Edmund Robinson, Helen Ronan,
Lyn Setchell, Jane Towle, Julie Wakeling, Sarah Williams, Dawn Wright.

Contributors:
Paddy Gannon, Gemma Hallam, Barbara Mascetti, Moira Steven, Paul Warren.

ISBN: 978 1 84146 641 5

With thanks to Karen Wells for the proofreading.
With thanks to Jan Greenway for the copyright research.

Graph to show trend in atmospheric CO_2 concentration and global temperature on page 37 based on data by EPICA Community Members 2004 and Siegenthaler et al 2005.

Data used to construct stopping distance diagram on page 47 From the Highway Code. © Crown Copyright re-produced under the terms of the Click-Use licence.

With thanks to iStockphoto.com for permission to use the images on page 90.

With thanks to Science Photo Library for permission to use the images on pages 127 and 128.

Every effort has been made to locate copyright holders and obtain permission to reproduce sources. For those sources where it has been difficult to trace the originator of the work, we would be grateful for information. If any copyright holder would like us to make an amendment to the acknowledgements, please notify us and we will gladly update the book at the next reprint. Thank you.

Groovy website: www.cgpbooks.co.uk

Printed by Elanders Ltd, Newcastle upon Tyne.
Jolly bits of clipart from CorelDRAW®

The Scientific Process

Before you get started with the really fun stuff, it's a good idea to understand exactly how the world of science works. Investigate these next few pages and you'll be laughing all day long on results day.

Scientists Come Up with Hypotheses — Then Test Them

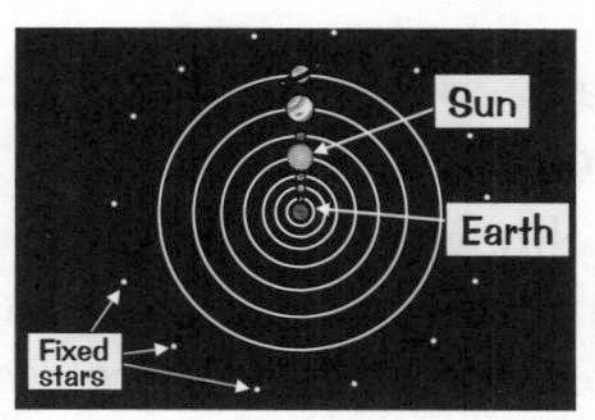

About 500 years ago, we still thought the Solar System looked like this.

1) Scientists try to explain things. Everything.
2) They start by observing or thinking about something they don't understand — it could be anything, e.g. planets in the sky, a person suffering from an illness, what matter is made of... anything.
3) Then, using what they already know (plus a bit of insight), they come up with a hypothesis — a possible explanation for what they've observed.
4) The next step is to test whether the hypothesis might be right or not — this involves gathering evidence (i.e. data from investigations).
5) To gather evidence the scientist uses the hypothesis to make a prediction — a statement based on the hypothesis that can be tested by carrying out experiments.
6) If the results from the experiments match the prediction, then the scientist can be more confident that the hypothesis is correct. This doesn't mean the hypothesis is true though — other predictions based on the hypothesis might turn out to be wrong.

Scientists Work Together to Test Hypotheses

1) Different scientists can look at the same evidence and interpret it in different ways. That's why scientists usually work in teams — they can share their different ideas on how to interpret the data they find.
2) Once a team has come up with (and tested) a hypothesis they all agree with, they'll present their work to the scientific community through journals and scientific conferences so it can be judged — this is called the peer review process.
3) Other scientists then check the team's results (by trying to replicate them) and carry out their own experiments to collect more evidence.
4) If all the experiments in the world back up the hypothesis, scientists start to have a lot of confidence in it. (A hypothesis that is accepted by pretty much every scientist is referred to as a theory.)
5) However, if another scientist does an experiment and the results don't fit with the hypothesis (and other scientists can replicate these results), then the hypothesis is in trouble. When this happens, scientists have to come up with a new hypothesis (maybe a modification of the old explanation, or maybe a completely new one).

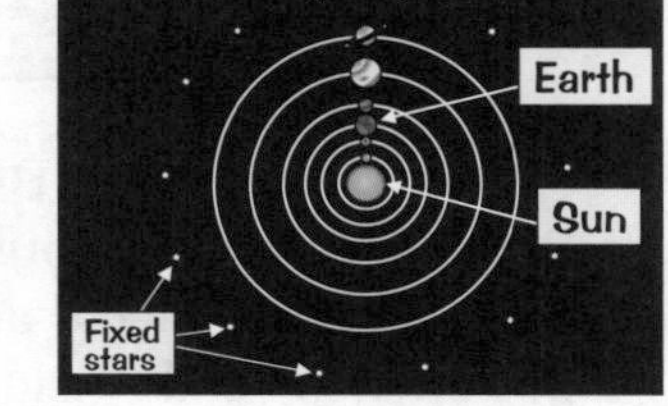

Then we thought it looked like this.

Scientific Ideas Change as New Evidence is Found

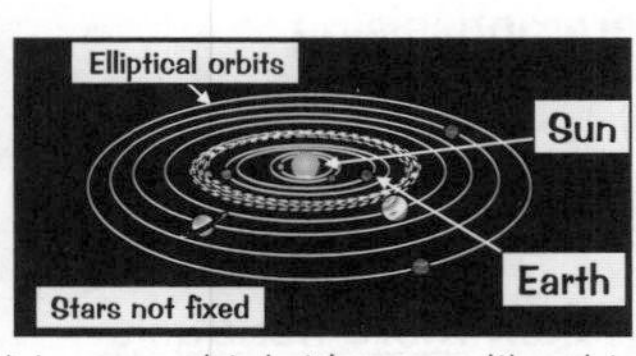

Now we think it's more like this.

1) Scientific explanations are provisional because they only explain the evidence that's currently available — new evidence may come up that can't be explained.
2) This means that scientific explanations never become hard and fast, totally indisputable fact. As new evidence is found (or new ways of interpreting existing evidence are found), hypotheses can change or be replaced.
3) Sometimes, an unexpected observation or result will suddenly throw a hypothesis into doubt and further experiments will need to be carried out. This can lead to new developments that increase our understanding of science.

You expect me to believe that — then show me the evidence...

If scientists think something is true, they need to produce evidence to convince others — it's all part of testing a hypothesis. One hypothesis might survive these tests, while others won't — it's how things progress. And along the way some hypotheses will be disproved — i.e. shown not to be true.

Quality of Data

Evidence is the key to science — but not all evidence is equally good.
The way evidence is gathered can have a big effect on how trustworthy it is...

The Bigger the Sample Size the Better

1) Data based on small samples isn't as good as data based on large samples. A sample should be representative of the whole population (i.e. it should share as many of the various characteristics in the population as possible) — a small sample can't do that as well.
2) The bigger the sample size the better, but scientists have to be realistic when choosing how big. For example, if you were studying how lifestyle affects people's weight it'd be great to study everyone in the UK (a huge sample), but it'd take ages and cost a bomb. Studying a thousand people is more realistic.

Evidence Needs to be Reliable (Repeatable and Reproducible)

Evidence is only reliable if it can be repeated (during an experiment) AND other scientists can reproduce it too (in other experiments). If it's not reliable, you can't believe it.

RELIABLE means that the data can be repeated, and reproduced by others.

EXAMPLE: In 1989, two scientists claimed that they'd produced 'cold fusion' (the energy source of the Sun — but without the big temperatures). It was huge news — if true, it would have meant cheap and abundant energy for the world... forever. However, other scientists just couldn't reproduce the results — so the results weren't reliable. And until they are, 'cold fusion' isn't going to be accepted as fact.

Evidence Also Needs to Be Valid

VALID means that the data is reliable AND answers the original question.

EXAMPLE: DO POWER LINES CAUSE CANCER?
Some studies have found that children who live near overhead power lines are more likely to develop cancer. What they'd actually found was a correlation (relationship) between the variables "presence of power lines" and "incidence of cancer" — they found that as one changed, so did the other. But this evidence is not enough to say that the power lines cause cancer, as other explanations might be possible. For example, power lines are often near busy roads, so the areas tested could contain different levels of pollution from traffic. So these studies don't show a definite link and so don't answer the original question.

Don't Always Believe What You're Being Told Straight Away

1) People who want to make a point might present data in a biased way, e.g. by overemphasising a relationship in the data. (Sometimes without knowing they're doing it.)
2) And there are all sorts of reasons why people might want to do this — for example, companies might want to 'big up' their products. Or make impressive safety claims.
3) If an investigation is done by a team of highly-regarded scientists it's sometimes taken more seriously than evidence from less well known scientists.
4) But having experience, authority or a fancy qualification doesn't necessarily mean the evidence is good — the only way to tell is to look at the evidence scientifically (e.g. is it reliable, valid, etc.).

RRRR — Remember, Reliable means Repeatable and Reproducible...

By now you should have realised how important trustworthy evidence is (even more important than a good supply of spot cream). Without it (the evidence, not the spot cream), you just can't believe what you're being told.

Limits of Science and the Issues it Creates

Science can give us amazing things — cures for diseases, space travel, heated toilet seats... But science has its limitations — there are questions that it just can't answer.

Some Questions Are Unanswered, Others are Unanswerable

1) Some questions are unanswered — we don't know everything and we never will. We'll find out more as new hypotheses are suggested and more experiments are done, but there'll always be stuff we don't know.
2) For example, we don't know what the exact impacts of global warming are going to be. At the moment scientists don't all agree on the answers because there isn't enough reliable and valid evidence.
3) Then there's the other type... questions that all the experiments in the world won't help us answer — the "Should we be doing this at all?" type questions. There are always two sides...
4) Take space exploration. It's possible to do it — but does that mean we should?
5) Different people have different opinions.

For example...
Some people say it's a good idea... it increases our knowledge about the Universe, we develop new technologies that can be useful on Earth too, it inspires young people to take an interest in science, etc.
Other people say it's a bad idea... the vast sums of money it costs should be spent on more urgent problems, like providing clean drinking water and curing diseases in poor countries. Others say that we should concentrate research efforts on understanding our own planet better first.

6) The question of whether something is morally or ethically right or wrong can't be answered by more experiments — there is no "right" or "wrong" answer.
7) The best we can do is get a consensus from society — a judgement that most people are more or less happy to live by. Science can provide more information to help people make this judgement, and the judgement might change over time. But in the end it's up to people and their conscience.

Scientific Developments are Great, but they can Raise Issues

Scientific knowledge is increased by doing experiments. And this knowledge leads to scientific developments, e.g. new technologies or new advice. These developments can create issues though. For example:

Economic issues: Society can't always afford to do things scientists recommend (e.g. investing heavily in alternative energy sources) without cutting back elsewhere.

Social issues: Decisions based on scientific evidence affect people — e.g. should fossil fuels be taxed more highly (to invest in alternative energy)? Should alcohol be banned (to prevent health problems)? Would the effect on people's lifestyles be acceptable...

Environmental issues: Nuclear power stations can provide us with a reliable source of electricity, but disposing of the waste can lead to environmental issues.

Ethical issues: There are a lot of things that scientific developments have made possible, but should we do them? E.g. develop better nuclear weapons.

Chips or rice? — totally unanswerable by science...

Science can't tell you whether you should or shouldn't do something. That is up to you and society to decide.

Planning Investigations

The next few pages show how investigations should be carried out — by both professional scientists and you.

To Make an Investigation a Fair Test You Have to Control the Variables

1) In a lab experiment you usually change one variable and measure how it affects the other variable.

> EXAMPLE: you might change only the angle of a slope and measure how it affects the time taken for a toy car to travel down it.

2) To make it a fair test everything else that could affect the results should stay the same (otherwise you can't tell if the thing that's being changed is affecting the results or not — the data won't be reliable or valid).

> EXAMPLE continued: you need to keep the slope length the same, otherwise you won't know if any change in the time taken is caused by the change in angle, or the change in length.

3) The variable that you change is called the independent variable.
4) The variable that's measured is called the dependent variable.
5) The variables that you keep the same are called control variables.
6) Because you can't always control all the variables, you often need to use a control experiment — an experiment that's kept under the same conditions as the rest of the investigation, but doesn't have anything done to it. This is so that you can see what happens when you don't change anything at all.

> EXAMPLE continued:
> Independent variable = angle of slope
> Dependent variable = time taken
> Control variable = length of slope

The Equipment Used has to be Right for the Job

Accurate data is data that's close to the true value — see the next page.

1) The measuring equipment you use has to be sensitive enough to accurately measure the chemicals you're using, e.g. if you need to measure out 11 ml of a liquid, you'll need to use a measuring cylinder that can measure to 1 ml, not 5 or 10 ml.
2) The smallest change a measuring instrument can detect is called its RESOLUTION. E.g. some mass balances have a resolution of 1 g and some have a resolution of 0.1 g.
3) Also, equipment needs to be calibrated so that your data is more accurate. E.g. mass balances need to be set to zero before you start weighing things.

Experiments Must be Safe

1) Part of planning an investigation is making sure that it's safe.
2) A hazard is something that can potentially cause harm.
3) There are lots of hazards you could be faced with during an investigation, e.g. radiation, electricity, gas, chemicals and fire.
4) You should always make sure that you identify all the hazards that you might encounter.
5) You should also come up with ways of reducing the risks from the hazards you've identified.
6) One way of doing this is to carry out a risk assessment:

For an experiment involving a Bunsen burner, the risk assessment might be something like this:

> Hazard: Bunsen burner is a fire risk.
> Precautions:
> - Keep flammable chemicals away from the Bunsen.
> - Never leave the Bunsen unattended when lit.
> - Always turn on the yellow safety flame when not in use.

Hazard: revision boredom. Precaution: use CGP books

Labs are dangerous places — you need to know the hazards of what you're doing before you start.

Collecting Data

There are a few things that can be done to make sure that you get the best results you possibly can.

Data Should be as Reliable, Accurate and Precise as Possible

1) When carrying out an investigation, you can improve the reliability of your results (see p. 2) by repeating the readings and calculating the mean (average, see next page). You should repeat readings at least twice (so that you have at least three readings to calculate an average result).
2) To make sure your results are reliable you can cross check them by taking a second set of readings with another instrument (or a different observer).
3) Checking your results match with secondary sources, e.g. studies that other people have done, also increases the reliability of your data.
4) You should always make sure that your results are accurate. Really accurate results are those that are really close to the true answer.

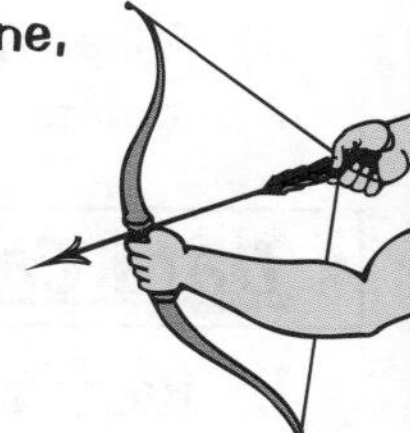

5) You can get accurate results by doing things like making sure the equipment you're using is sensitive enough (see previous page), and by recording your data to a suitable level of accuracy. For example, if you're taking digital readings of something, the results will be more accurate if you include at least a couple of decimal places instead of rounding to whole numbers.
6) You should also always make sure your results are precise. Precise results are ones where the data is all really close to the mean (i.e. not spread out).

Trial Runs Help Figure out the Range and Interval of Variable Values

1) Before you carry out an experiment, it's a good idea to do a trial run first — a quick version of your experiment.
2) Trial runs help you work out whether your plan is right or not — you might decide to make some changes after trying out your method.
3) Trial runs are used to figure out the range of variable values used (the upper and lower limit).
4) And they're used to figure out the interval (gaps) between the values too.
5) Trial runs can also help you figure out how many times the experiment has to be repeated to get reliable results. E.g. if you repeat it two times and the results are all similar, then two repeats is enough.

Slope example from previous page continued:

- You might do trial runs at 20, 40, 60 and 80°. If the time taken is too short to accurately measure at 80°, you might narrow the range to 20-60°.
- If using 20° intervals gives you a big change in time taken you might decide to use 10° intervals, e.g. 20, 30, 40, 50°...

You Can Check For Mistakes Made When Collecting Data

1) When you've collected all the results for an experiment, you should have a look to see if there are any results that don't seem to fit in with the rest.
2) Most results vary a bit, but any that are totally different are called anomalous results.
3) They're caused by human errors, e.g. by a whoopsie when measuring.
4) The only way to stop them happening is by taking all your measurements as carefully as possible.
5) If you ever get any anomalous results, you should investigate them to try to work out what happened. If you can work out what happened (e.g. you measured something wrong) you can ignore them when processing your results.

Reliable data — it won't ever forget your birthday...

All this stuff is really important — without good quality data an investigation will be totally meaningless. So give this page a read through a couple of times and your data will be the envy of the whole scientific community.

Processing, Presenting and Interpreting Data

The fun doesn't stop once the data's been collected — it then needs to be processed and presented...

Data Needs to be Organised

1) Data that's been collected needs to be organised so it can be processed later on.
2) Tables are dead useful for organising data.
3) When drawing tables you should always make sure that each column has a heading and that you've included the units.
4) Annoyingly, tables are about as useful as a chocolate teapot for showing patterns or relationships in data. You need to use some kind of graph or mathematical technique for that...

Test tube	Result (ml)	Repeat 1 (ml)	Repeat 2 (ml)
A	28	37	32
B	47	51	60
C	68	72	70

Data Can be Processed Using a Bit of Maths

1) Raw data generally just ain't that useful. You usually have to process it in some way.
2) A couple of the most simple calculations you can perform are the mean (average) and the range (how spread out the data is):

- To calculate the mean ADD TOGETHER all the data values and DIVIDE by the total number of values. You usually do this to get a single value from several repeats of your experiment.
- To calculate the range find the LARGEST number and SUBTRACT the SMALLEST number. You usually do this to check the accuracy and reliability of the results — the greater the spread of the data, the lower the accuracy and reliability.

Test tube	Result (ml)	Repeat 1 (ml)	Repeat 2 (ml)	Mean (ml)	Range
A	28	37	32	(28 + 37 + 32) ÷ 3 = 32.3	37 – 28 = 9
B	47	51	60	(47 + 51 + 60) ÷ 3 = 52.7	60 – 47 = 13
C	68	72	70	(68 + 72 + 70) ÷ 3 = 70.0	72 – 68 = 4

Different Types of Data Should be Presented in Different Ways

1) Once you've carried out an investigation, you'll need to present your data so that it's easier to see patterns and relationships in the data.
2) Different types of investigations give you different types of data, so you'll always have to choose what the best way to present your data is.

Pie charts can be used to present the same sort of data as bar charts. They're mostly used when the data is in percentages or fractions though.

Bar Charts

If the independent variable is categoric (comes in distinct categories, e.g. blood types, metals) you should use a bar chart to display the data. You also use them if the independent variable is discrete (the data can be counted in chunks, where there's no in-between value, e.g. number of people is discrete because you can't have half a person).

There are some golden rules you need to follow for drawing bar charts:

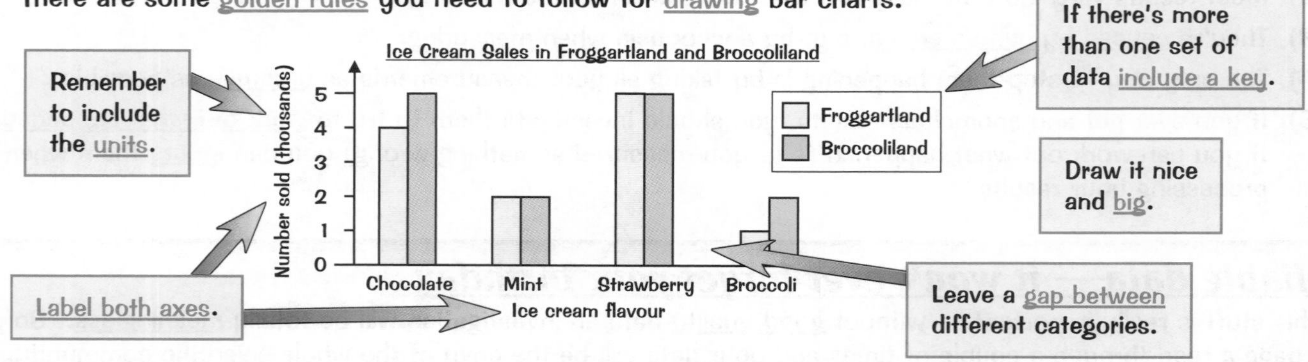

Processing, Presenting and Interpreting Data

Line Graphs

If the independent variable is continuous (numerical data that can have any value within a range, e.g. length, volume, temperature) you should use a line graph to display the data.

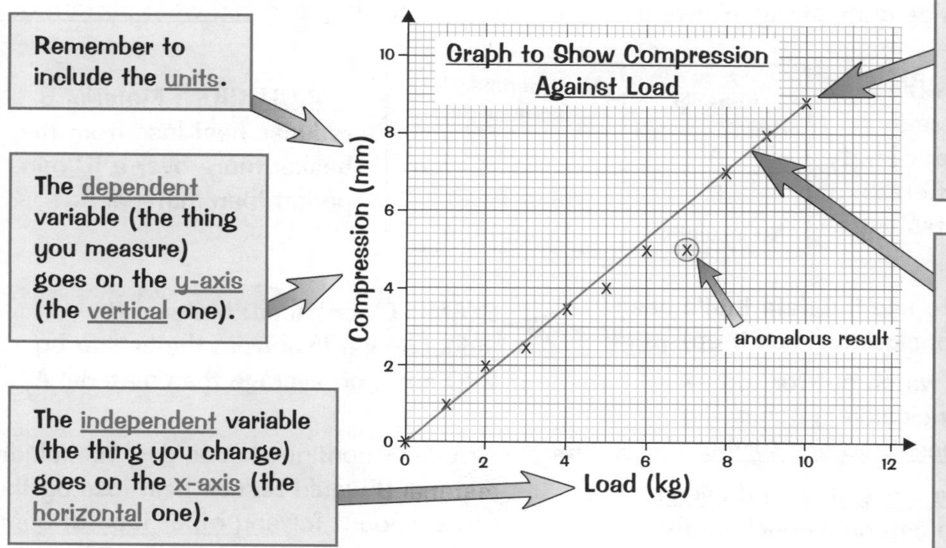

When plotting points, use a sharp pencil and make a neat little cross (don't do blobs).

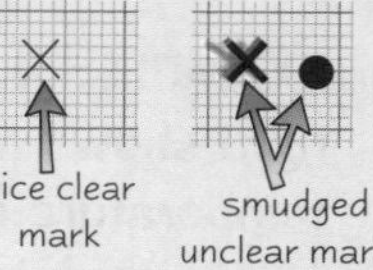

Don't join the dots up. You should draw a line of best fit (or a curve of best fit if your points make a curve). When drawing a line (or curve), try to draw the line through or as near to as many points as possible, ignoring anomalous results.

Line Graphs Can Show Relationships in Data

1) Line graphs are great for showing relationships between two variables (just like other graphs).
2) Here are some of the different types of correlation (relationship) shown on line graphs:

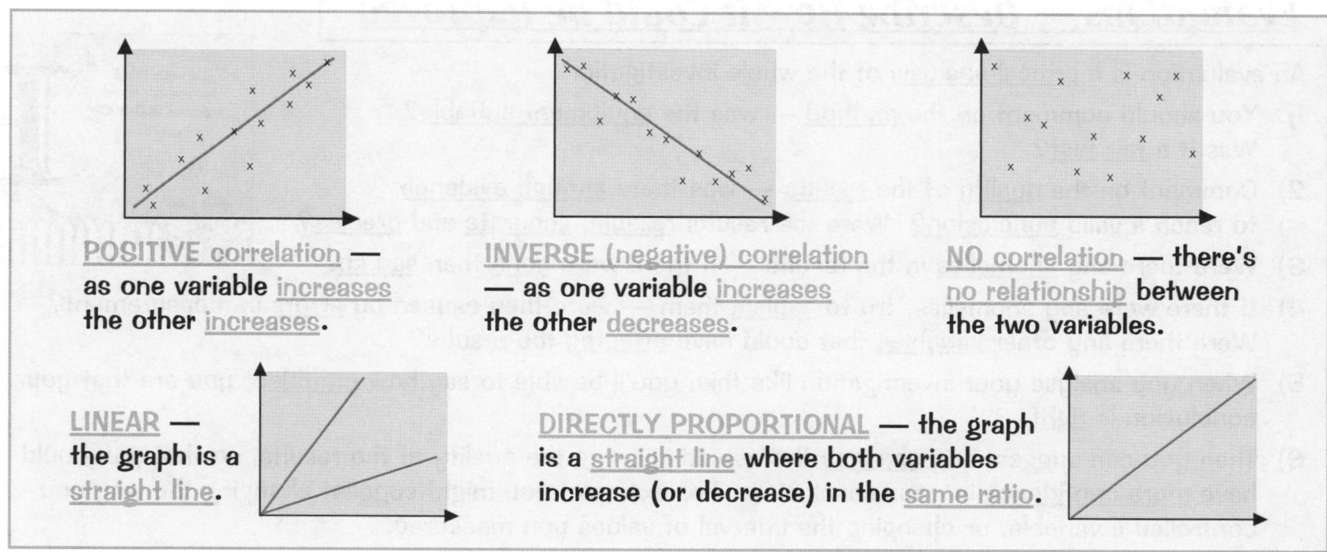

3) You've got to be careful not to confuse correlation with cause though. A correlation just means that there's a relationship between two variables. It doesn't always mean that the change in one variable is causing the change in the other.
4) There are three possible reasons for a correlation. It could be down to chance, it could be that there's a third variable linking the two things, or it might actually be that one variable is causing the other to change.

There's a positive correlation between age of man and length of nose hair...

Process, present, interpret... data's like a difficult child — it needs a lot of attention. Go on, make it happy.

Concluding and Evaluating

At the end of an investigation, the conclusion and evaluation are waiting. Don't worry, they won't bite.

A Conclusion is a Summary of What You've Learnt

1) Once all the data's been collected, presented and analysed, an investigation will always involve coming to a conclusion.
2) Drawing a conclusion can be quite straightforward — just look at your data and say what pattern you see.

EXAMPLE: The table on the right shows the decrease in temperature of a beaker of hot water insulated with different materials over 10 minutes.

Material	Mean temperature decrease (°C)
A	4
B	2
No insulation	20

CONCLUSION: Material B reduces heat loss from the beaker more over a 10 minute period than material A.

3) However, you also need to use the data that's been collected to justify the conclusion (back it up).

EXAMPLE continued: Material B reduced heat loss from the beaker by 2 °C more on average than material A.

4) There are some things to watch out for too — it's important that the conclusion matches the data it's based on and doesn't go any further.

EXAMPLE continued: You can't conclude that material B would reduce heat loss by the same amount for any other type of container — the results could be totally different.

5) Remember not to confuse correlation and cause (see previous page). You can only conclude that one variable is causing a change in another if you have controlled all the other variables (made it a fair test).

Evaluations — Describe How it Could be Improved

An evaluation is a critical analysis of the whole investigation.

1) You should comment on the method — was the equipment suitable? Was it a fair test?
2) Comment on the quality of the results — was there enough evidence to reach a valid conclusion? Were the results reliable, accurate and precise?
3) Were there any anomalies in the results — if there were none then say so.
4) If there were any anomalies, try to explain them — were they caused by errors in measurement? Were there any other variables that could have affected the results?
5) When you analyse your investigation like this, you'll be able to say how confident you are that your conclusion is right.
6) Then you can suggest any changes that would improve the quality of the results, so that you could have more confidence in your conclusion. For example, you might suggest changing the way you controlled a variable, or changing the interval of values you measured.
7) You could also make more predictions based on your conclusion, then further experiments could be carried out to test them.
8) When suggesting improvements to the investigation, always make sure that you say why you think this would make the results better.

Evaluation — in my next study I will make sure I don't burn the lab down...

I know it doesn't seem very nice, but writing about where you went wrong is an important skill — it shows you've got a really good understanding of what the investigation was about. It's difficult for me — I'm always right.

Heat, Temperature and Kinetic Theory

When it starts to get a bit nippy, on goes the heating to warm things up a bit.
Heating is all about the transfer of energy. Here are a few useful definitions to begin with.

Heat is a Measure of Energy

1) When a substance is heated, its particles gain kinetic energy (KE). This energy makes the particles in a gas or a liquid move around faster. In a solid, the particles vibrate more rapidly (see below).
2) This energy is measured on an absolute scale. (This means it can't go lower than zero, because there's a limit to how slow particles can move.) The unit of heat energy is the joule (J).

Temperature is a Measure of Hotness

1) Temperature is a measure of the average kinetic energy of the particles in a substance. The hotter something is, the higher its temperature, and the higher the average KE of its particles.
2) Temperature is usually measured in °C (degrees Celsius), but there are other temperature scales, like °F (degrees Fahrenheit). These are not absolute scales as they can go below zero.

Energy tends to flow from hot objects to cooler ones. E.g. warm radiators heat the cold air in your room — they'd be no use if heat didn't flow.

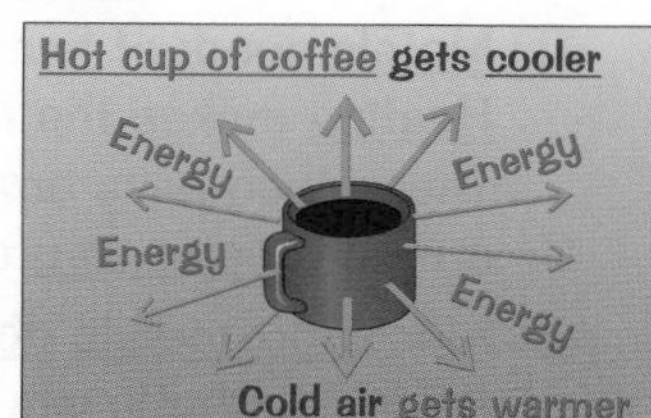

If there's a DIFFERENCE IN TEMPERATURE between two places, then ENERGY WILL FLOW between them.

The greater the difference in temperature, the faster the rate of cooling will be.
E.g. a hot cup of coffee will cool down quicker in a cold room than in a warm room.

Kinetic Theory Can Explain the Three States of Matter

The three states of matter are solid (e.g. ice), liquid (e.g. water) and gas (e.g. water vapour). The particles of a particular substance in each state are the same — only the arrangement and energy of the particles are different.

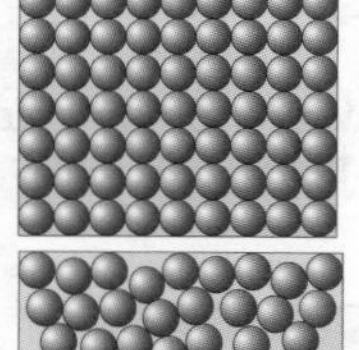

SOLIDS — strong forces of attraction hold the particles close together in a fixed, regular arrangement. The particles don't have much energy so they can only vibrate about their fixed positions.

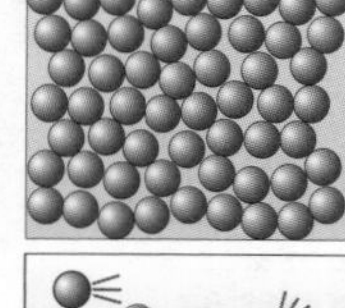

LIQUIDS — there are weaker forces of attraction between the particles. The particles are close together, but can move past each other, and form irregular arrangements. They have more energy than the particles in a solid — they move in random directions at low speeds.

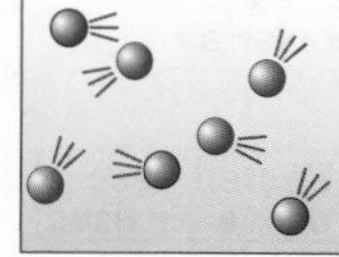

GASES — There are almost no forces of attraction between the particles. The particles have more energy than those in liquids and solids — they are free to move, and travel in random directions and at high speeds.

When you heat a substance, you give its particles more kinetic energy (KE) — they vibrate or move faster. This is what eventually causes solids to melt and liquids to boil.

Moving in random directions at high speeds? Sounds like my dancing...

Whether a material is a solid, liquid or a gas depends on how much energy the particles have.
For some of the lighter elements, you have to make it pretty darn chilly before they even think about condensing (see p.12) from a gas to a liquid — helium for example doesn't condense until it's cooled to 4.2 K (-268.8 °C).

Conduction and Convection

Particles in solids can't flow, but sitting close together in regular patterns makes them good at conduction. Gases and liquids are usually free to slosh about — and that allows them to transfer heat by convection.

Conduction Occurs Mainly in Solids

Houses lose a lot of heat through their windows even when they're shut. Heat flows from the warm inside face of the window to the cold outside face mostly by conduction.

1) In a solid, the particles are held tightly together. So when one particle vibrates, it bumps into other particles nearby and quickly passes the vibrations on.
2) Particles which vibrate faster than others pass on their extra kinetic energy to neighbouring particles. These particles then vibrate faster themselves.

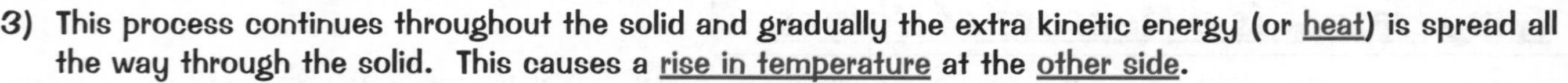

3) This process continues throughout the solid and gradually the extra kinetic energy (or heat) is spread all the way through the solid. This causes a rise in temperature at the other side.

CONDUCTION OF HEAT is the process where vibrating particles pass on extra kinetic energy to neighbouring particles.

4) Metals conduct heat really well because some of their electrons are free to move inside the metal. Heating makes the electrons move faster and collide with other free electrons, transferring energy. These then pass on their extra energy to other electrons, etc. Because the electrons move freely, this is a much faster way of transferring energy than slowly passing it between jostling neighbouring atoms.
5) Most non-metals don't have free electrons, so warm up more slowly, making them good for insulating things — that's why metals are used for saucepans, but non-metals are used for saucepan handles.
6) Liquids and gases conduct heat more slowly than solids — the particles aren't held so tightly together, which prevents them bumping into each other so often. So air is a good insulator.

Convection Occurs in Liquids and Gases

1) When you heat up a liquid or gas, the particles move faster, and the fluid (liquid or gas) expands, becoming less dense.
2) The warmer, less dense fluid rises above its colder, denser surroundings, like a hot air balloon does.
3) As the warm fluid rises, cooler fluid takes its place. As this process continues, you actually end up with a circulation of fluid (convection currents). This is how immersion heaters work.

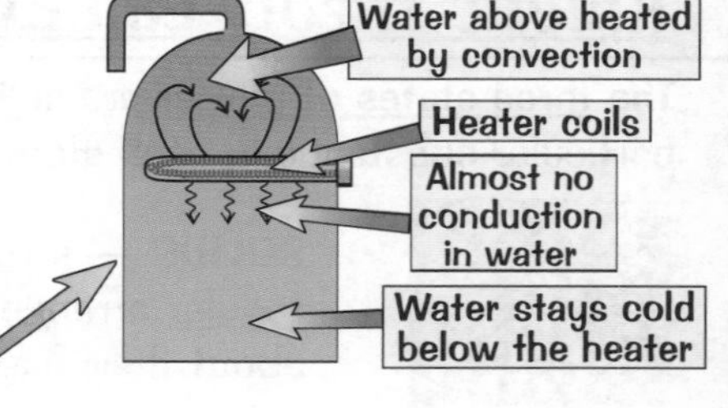

CONVECTION occurs when the more energetic particles move from the hotter region to the cooler region — and take their heat energy with them.

4) Radiators in the home rely on convection to make the warm air circulate round the room.
5) Convection can't happen in solids because the particles can't move — they just vibrate on the spot.
6) To reduce convection, you need to stop the fluid moving. Clothes, blankets and cavity wall foam insulation all work by trapping pockets of air. The air can't move so the heat has to conduct very slowly through the pockets of air, as well as the material in between.

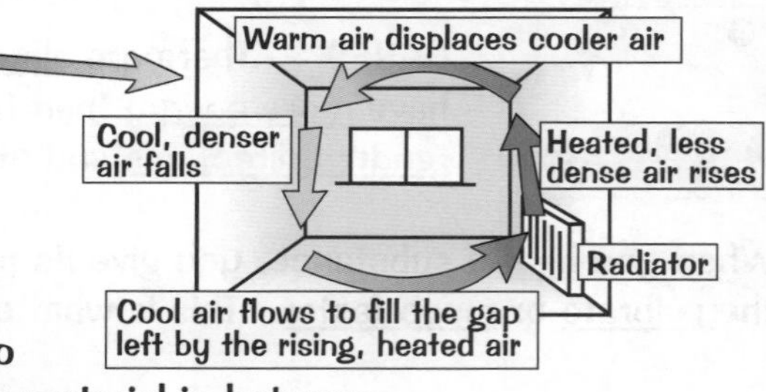

The good old garden spade is a great example of conduction...

If a garden spade is left outside in cold weather, the metal bit will always feel colder than the wooden handle. But it isn't colder — it just conducts heat away from your hand quicker. The opposite is true if the spade is left out in the sunshine — it'll feel hotter because it conducts heat into your hand quicker.

Heat Radiation

Be careful not to get confused between heat radiation and ionising radiation (see page 72). This page is all about heat energy and the types of materials that are good at absorbing and emitting it.

Thermal Radiation Involves Emission of Electromagnetic Waves

Heat radiation consists purely of electromagnetic waves of a certain range of frequencies — infrared radiation. It's next to visible light in the electromagnetic spectrum (see p.32).

> Radiation doesn't need a medium (material) to travel through, so it can occur in a vacuum, like space. This is the only way that heat reaches us from the Sun.

1) All objects are continually emitting and absorbing heat radiation.
2) An object that's hotter than its surroundings emits more radiation than it absorbs (as it cools down). And an object that's cooler than its surroundings absorbs more radiation than it emits (as it warms up).
3) The hotter an object is, the more radiation it radiates in a given time.
4) Power is the just the rate of energy change — that's energy ÷ time (see p.49). For an object to stay at the same temperature, the power of heat absorbed needs to be the same as the power emitted.
5) You can feel this heat radiation if you stand near something hot like a fire or if you put your hand just above the bonnet of a recently parked car.

(recently parked car)

(after an hour or so)

Radiation Depends an Awful Lot on Surface Colour and Texture

1) Dark matt surfaces absorb heat radiation falling on them much better than light glossy surfaces, such as gloss white or silver. They also emit much more heat radiation (at any given temperature).
2) Silvered surfaces reflect nearly all heat radiation falling on them.

Solar hot water panels

1) Solar hot water panels contain water pipes under a black surface (or black painted pipes under glass).
2) Heat radiation from the Sun is absorbed by the black surface to heat the water in the pipes.

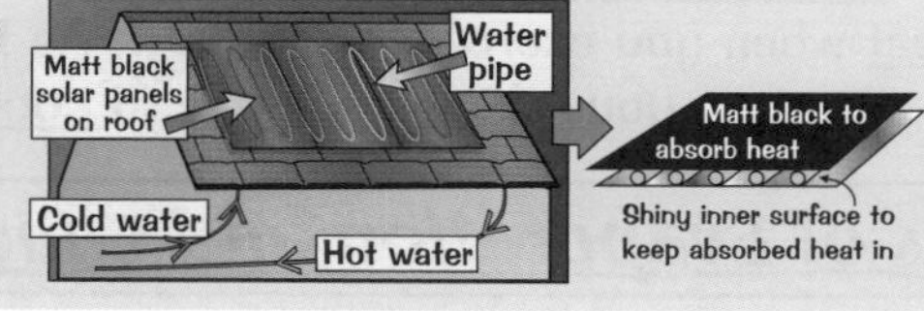

Survival blankets

1) If someone gets injured halfway up a big snowy hill, it can be crucial to keep them as warm as possible till help arrives.
2) A silver coloured blanket helps to stop their body heat radiating away — and could save their life.

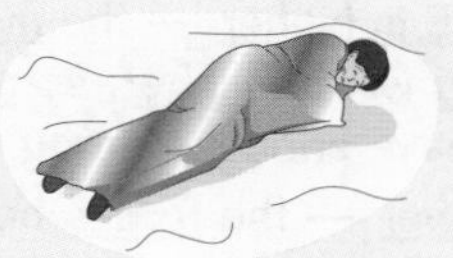

There Are Lots of Fairly Dull Experiments to Demonstrate This...

Here are two of the most gripping:

Leslie's Cube

The matt black side emits most heat, so it's that thermometer which gets hottest.

The Melting Wax Trick

The matt black surface absorbs most heat, so its wax melts first and the ball bearing drops.

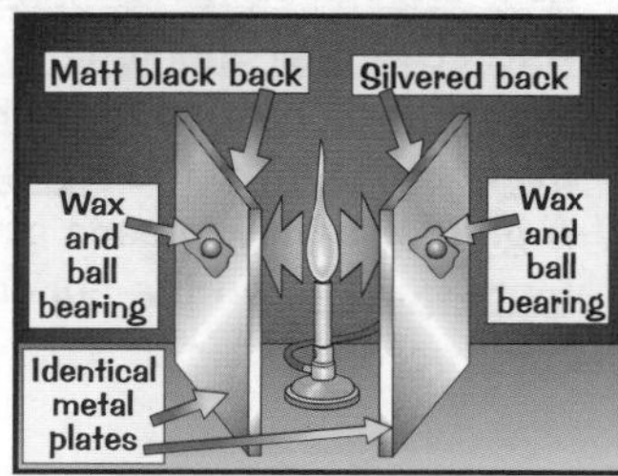

I know it's Leslie's Cube — but he said I could borrow it...

The key idea here is that heat radiation is affected by the colour and texture of surfaces. Thermal radiation questions often ask you why something's painted silver, or how you could reduce the heat losses from something.

Condensation and Evaporation

Here are a couple more things about particles in gases and liquids you need to think about. It's party-cle time...

Condensation is When Gas Turns to Liquid

1) When a gas cools, the particles in the gas slow down and lose kinetic energy. The attractive forces between the particles pull them closer together.
2) If the temperature gets cold enough and the gas particles get close enough together that condensation can take place, the gas becomes a liquid.
3) Water vapour in the air condenses when it comes into contact with cold surfaces e.g. drinks glasses.
4) The steam you see rising from a boiling kettle is actually invisible water vapour condensing to form tiny water droplets as it spreads into cooler air.

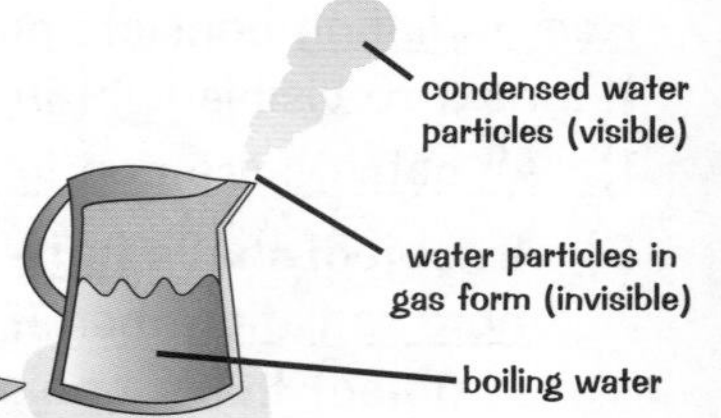

Evaporation is When Liquid Turns to Gas

1) Evaporation is when particles escape from a liquid.
2) Particles can evaporate from a liquid at temperatures that are much lower than the liquid's boiling point.
3) Particles near the surface of a liquid can escape and become gas particles if:

- The particles are travelling in the right direction to escape the liquid.
- The particles are travelling fast enough (they have enough kinetic energy) to overcome the attractive forces of the other particles in the liquid.

4) The fastest particles (with the most kinetic energy) are most likely to evaporate from the liquid — so when they do, the average speed and kinetic energy of the remaining particles decreases.
5) This decrease in average particle energy means the temperature of the remaining liquid falls — the liquid cools.
6) This cooling effect can be really useful. For example, you sweat when you exercise or get hot. As the water from the sweat on your skin evaporates, it cools you down.

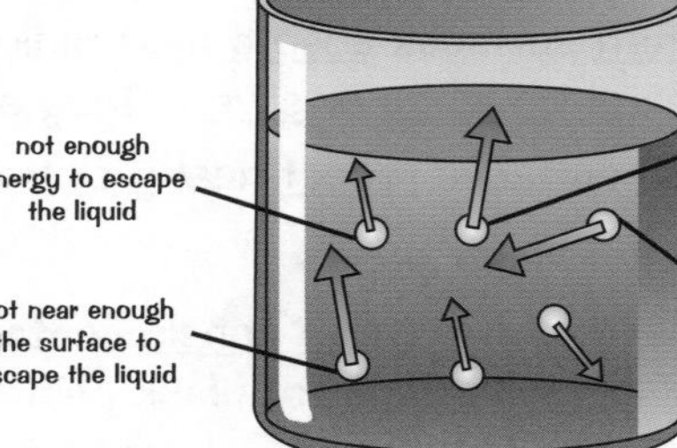

Rates of Evaporation and Condensation can Vary

The RATE OF EVAPORATION will be faster if the...

- TEMPERATURE is higher — the average particle energy will be higher, so more particles will have enough energy to escape.
- DENSITY is lower — the forces between the particles will usually be weaker, so more particles will have enough energy to overcome these forces and escape the liquid.
- SURFACE AREA is larger — more particles will be near enough to the surface to escape the liquid.
- AIRFLOW over the liquid is greater — the lower the concentration of an evaporating substance in the air it's evaporating into, the higher the rate of evaporation. A greater airflow means air above the liquid is replaced more quickly, so the concentration in the air will be lower.

The RATE OF CONDENSATION will be faster if the...

- TEMPERATURE OF THE GAS is lower — the average particle energy in the gas is lower — so more particles will slow down enough to clump together and form liquid droplets.
- TEMPERATURE OF THE SURFACE THE GAS TOUCHES is lower.
- DENSITY is higher — the forces between the particles will be stronger. Fewer particles will have enough energy to overcome these forces and will instead clump together and form a liquid.
- AIRFLOW is less — the concentration of the substance in the air will be higher, and so the rate of condensation will be greater.

A little less condensation, a little more action...

The people who make adverts for drinks know what customers like to see — condensation on the outside of the bottle. It makes the drink look nice and cold and extra-refreshing. Mmmm. If it wasn't for condensation, you'd never be able to draw pictures on the bus window with your finger either — you've got a lot to be thankful for...

Rate of Heat Transfer

There are loads of factors that affect the rate of heat transfer.
Different objects can lose or gain heat much faster than others — even in the same conditions. Read on...

The Rate of Heat Energy Transfer Depends on Many Things...

1) Heat energy is radiated from the surface of an object.
2) The bigger the surface area, the more infrared waves that can be emitted from (or absorbed by) the surface — so the quicker the transfer of heat. E.g. radiators have large surface areas to maximise the amount of heat they transfer.
3) This is why car and motorbike engines often have 'fins' — they increase the surface area so heat is radiated away quicker. So the engine cools quicker.
4) Heat sinks are devices designed to transfer heat away from objects they're in contact with, e.g. computer components. They have fins and a large surface area so they can emit heat as quickly as possible.

Cooling fins on engines increase surface area to speed up cooling.

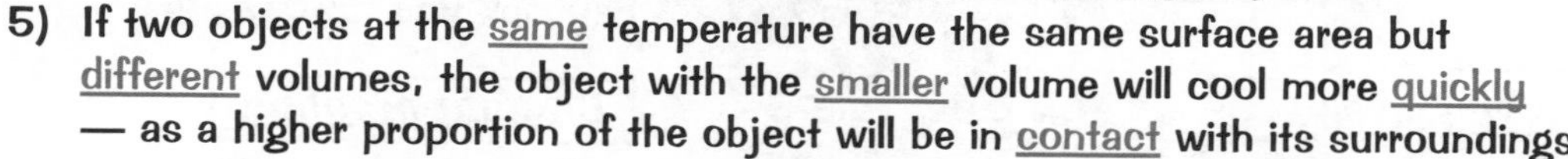

5) If two objects at the same temperature have the same surface area but different volumes, the object with the smaller volume will cool more quickly — as a higher proportion of the object will be in contact with its surroundings.
6) Other factors, like the type of material, affect the rate too. Objects made from good conductors (see p.10) transfer heat away more quickly than insulating materials, e.g. plastic. It also matters whether the materials in contact with it are insulators or conductors. If an object is in contact with a conductor, the heat will be conducted away much faster than if it is in contact with a good insulator.

Some Devices are Designed to Limit Heat Transfer

You need to know about heat energy transfers and how products can be designed to reduce them.

Vacuum Flasks

1) The glass bottle is double-walled with a vacuum between the two walls. This stops all conduction and convection through the sides.
2) The walls either side of the vacuum are silvered to keep heat loss by radiation to a minimum.
3) The bottle is supported using insulating foam. This minimises heat conduction to or from the outer glass bottle.
4) The stopper is made of plastic and filled with cork or foam to reduce any heat conduction through it.

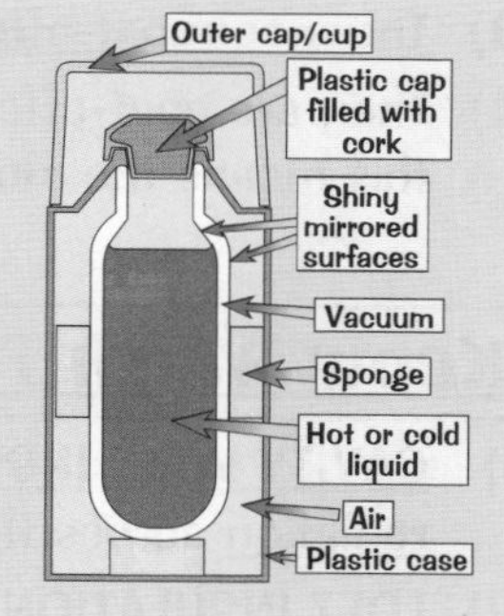

Humans and Animals Have Ways of Controlling Heat Transfer Too

1) In the cold, the hairs on your skin 'stand up' to trap a thicker layer of insulating air around the body. This limits the amount of heat loss by convection. Some animals do the same using fur.
2) When you're too warm, your body diverts more blood to flow near the surface of your skin so that more heat can be lost by radiation — that's why some people go pink when they get hot.
3) Generally, animals in warm climates have larger ears than those in cold climates to help control heat transfer.

For example, Arctic foxes have evolved small ears, with a small surface area to minimise heat loss by radiation and conserve body heat.
Desert foxes on the other hand have huge ears with a large surface area to allow them to lose heat by radiation easily and keep cool.

Don't call me 'Big Ears' — call me 'Large Surface Area'...

Examiners are like small children — they ask some barmy questions. If they ask you one about heat transfer, you must always say which form of heat transfer is involved at any point, either conduction, convection or radiation. You've got to show them that you know your stuff — it's the only way to get top marks.

Energy Efficiency in the Home

There are lots of things you can do to a building to reduce the amount of heat energy that escapes. Some are more effective than others, and some are better for your pocket than others. The most obvious examples are in the home, but you could apply this to any situation where you're trying to cut down energy loss.

Effectiveness and Cost-effectiveness are Not the Same...

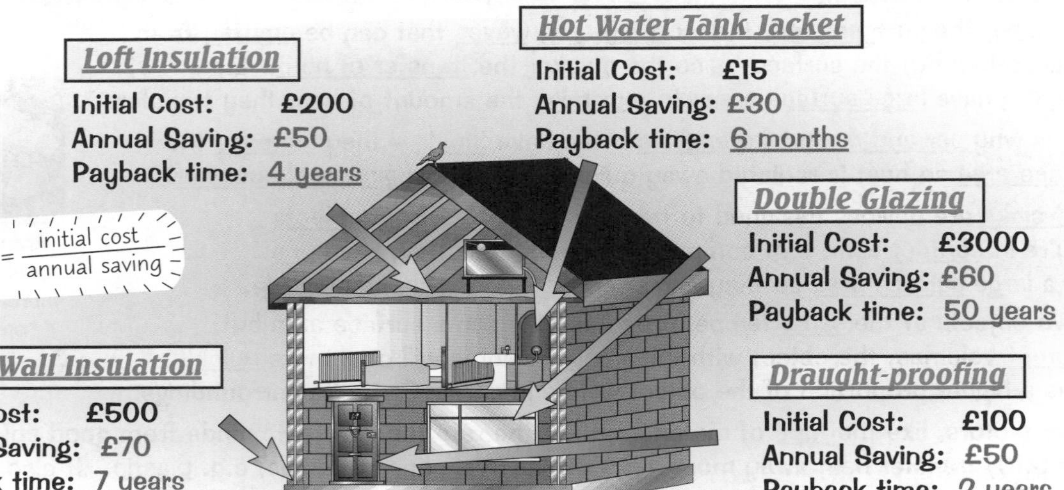

1) The most effective methods of insulation are ones that give you the biggest annual saving (they save you the most money each year on your heating bills).
2) Eventually, the money you've saved on heating bills will equal the initial cost of putting in the insulation (the amount it cost to buy). The time it takes is called the payback time.
3) The most cost-effective methods tend to be the cheapest.
4) They are cost-effective because they have a short payback time — this means the money you save covers the amount you paid really quickly.

Know Which Types of Heat Transfer Are Involved

1) CAVITY WALL INSULATION — foam squirted into the gap between the bricks reduces convection and radiation across the gap. Pockets of air in the foam reduce heat transfer by conduction.
2) LOFT INSULATION — a thick layer of fibreglass wool laid out across the whole loft floor reduces conduction and radiation into the roof space from the ceiling.
3) DRAUGHT-PROOFING — strips of foam and plastic around doors and windows stop draughts of cold air blowing in, i.e. they reduce heat loss due to convection.
4) HOT WATER TANK JACKET — lagging such as fibreglass wool reduces conduction and radiation.
5) THICK CURTAINS — big bits of cloth over the window to reduce heat loss by conduction and radiation.

U-Values Show How Fast Heat can Transfer Through a Material

1) Heat transfers faster through materials with higher U-values than through materials with low U-values.
2) So the better the insulator (see p.10) the lower the U-value. E.g. the U-value of a typical duvet is about 0.75 W/m^{2}K, whereas the U-value of loft insulation material is around 0.15 W/m^{2}K.

It's payback time...

And it's the same with, say, cars. Buying a more fuel-efficient car might sound like a great idea — but if it costs loads more than a clapped-out old fuel-guzzler, you might still end up out of pocket. If it's cost-effectiveness you're thinking about, you always have to offset initial cost against annual savings.

Specific Heat Capacity

Specific heat capacity is one of those topics that puts people off just because it has a weird name. If you can get over that, it's actually not too bad — it sounds a lot harder than it is. Go on. Give it a second chance.

Specific Heat Capacity Tells You How Much Energy Stuff Can Store

1) It takes more heat energy to increase the temperature of some materials than others. E.g. you need 4200 J to warm 1 kg of water by 1 °C, but only 139 J to warm 1 kg of mercury by 1 °C.
2) Materials which need to gain lots of energy to warm up also release loads of energy when they cool down again. They can 'store' a lot of heat.
3) The measure of how much energy a substance can store is called its specific heat capacity.
4) Specific heat capacity is the amount of energy needed to raise the temperature of 1 kg of a substance by 1 °C. Water has a specific heat capacity of 4200 J/kg°C.

There's a Handy Formula for Specific Heat Capacity

You'll have to do calculations involving specific heat capacity. This is the equation to learn:

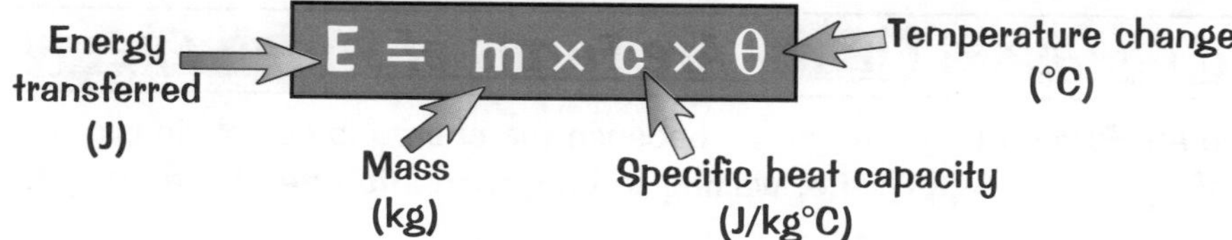

EXAMPLE: How much energy is needed to heat 2 kg of water from 10 °C to 100 °C?

ANSWER: Energy needed = $2 \times 4200 \times 90$ = 756 000 J

If you're not working out the energy, you'll have to rearrange the equation, so this formula triangle will come in dead handy. You cover up the thing you're trying to find. The parts of the formula you can still see are what it's equal to.

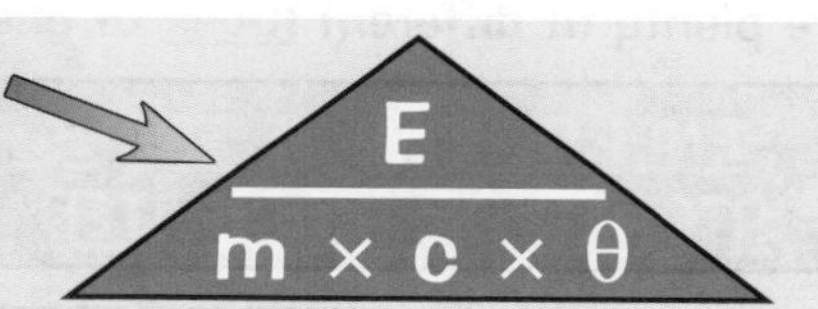

EXAMPLE: An empty 200 g aluminium kettle cools down from 115 °C to 10 °C, losing 19 068 J of heat energy. What is the specific heat capacity of aluminium?

Remember — you need to convert the mass to kilograms first.

ANSWER: $\text{SHC} = \frac{\text{Energy}}{\text{Mass} \times \text{Temp Ch}} = \frac{19\,068}{0.2 \times 105}$ = 908 J/kg°C

Heaters Have High Heat Capacities to Store Lots of Energy

1) The materials used in heaters usually have high specific heat capacities so that they can store large amounts of heat energy.
2) Water has a really high specific heat capacity. It's also a liquid, so it can easily be pumped around in pipes — ideal for central heating systems in buildings.
3) Electric storage heaters are designed to store heat energy at night (when electricity is cheaper), and then release it during the day. They store the heat using concrete or bricks, which (surprise surprise) have a high specific heat capacity (around 880 J/kg°C).
4) Some heaters are filled with oil, which has a specific heat capacity of around 2000 J/kg°C. Because this is lower than water's specific heat capacity, oil heating systems are often not as good as water-based systems. Oil does have a higher boiling point though, which usually means oil-filled heaters can safely reach higher temperatures than water-based ones.

I've just eaten five sausages — I have a high specific meat capacity...

I'm sure you'll agree that this isn't the most exciting part of physics — it's not about space travel, crashing cars or even using springs — but it might come up in your exams. If you knuckle down and get that formula triangle learnt then you'll be well on the way to breezing through this question, should it appear.

Energy Transfer

Heat is just one type of energy, but there are lots more as well:

The Nine Types of Energy

You should know all of these well enough by now to list them from memory, including the examples:

1) ELECTRICAL Energy.................................. — whenever a current flows.
2) LIGHT Energy.. — from the Sun, light bulbs, etc.
3) SOUND Energy.. — from loudspeakers or anything noisy.
4) KINETIC Energy, or MOVEMENT Energy......... — anything that's moving has it.
5) NUCLEAR Energy...................................... — released only from nuclear reactions.
6) THERMAL Energy or HEAT Energy............... — flows from hot objects to colder ones.
7) GRAVITATIONAL POTENTIAL Energy.............. — possessed by anything which can fall.
8) ELASTIC POTENTIAL Energy......................... — stretched springs, elastic, rubber bands, etc.
9) CHEMICAL Energy....................................... — possessed by foods, fuels, batteries etc.

Potential and Chemical Energy Are Forms of Stored Energy

The last three above are forms of stored energy because the energy is not obviously doing anything, it's kind of waiting to happen, i.e. waiting to be turned into one of the other forms.

There is a Conservation of Energy Principle

There are plenty of different types of energy, but they all obey the principle below:

ENERGY CAN BE TRANSFERRED USEFULLY FROM ONE FORM TO ANOTHER, STORED OR DISSIPATED — BUT IT CAN NEVER BE CREATED OR DESTROYED.

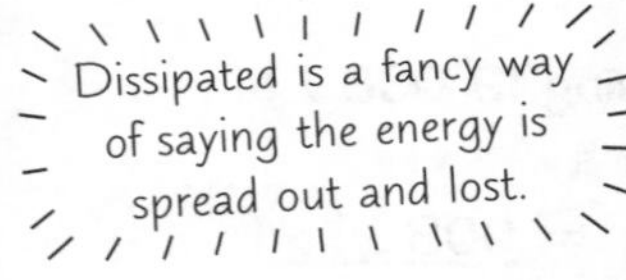

Gravitational Potential → Kinetic

Another important principle when learning about energy is:

Energy is only useful when it can be converted from one form to another.

Energy Transfers Show... well... the Transfer of Energy

In the exam, they might ask you about any device or energy transfer system they feel like. If you understand a few different examples, it'll be easier to think through whatever they ask you about in the exam.

EXAMPLES:

Electrical Devices, e.g. televisions: Electrical energy → Light, sound and heat energy
Batteries: Chemical energy → Electrical and heat energy
Electrical Generation, e.g. wind turbines: Kinetic energy → Electrical and heat energy
Potential Energy, e.g. firing a bow and arrow: Elastic potential energy → Kinetic and heat energy

Energy can't be created or destroyed — only talked about a lot...

Chemical energy → kinetic energy → electrical energy → kinetic energy → chemical energy.
(me thinking) (me typing) (my computer) (printing machine) (you reading this)

Efficiency of Machines

An open fire looks cosy, but a lot of its heat energy goes straight up the chimney, by convection, instead of heating up your living room. All this energy is 'wasted', so open fires aren't very efficient.

Machines Always Waste Some Energy

1) Useful machines are only useful because they convert energy from one form to another. Take cars for instance — you put in chemical energy (petrol or diesel) and the engine converts it into kinetic (movement) energy.
2) The total energy output is always the same as the energy input, but only some of the output energy is useful. So for every joule of chemical energy you put into your car you'll only get a fraction of it converted into useful kinetic energy.
3) This is because some of the input energy is always lost or wasted, often as heat. In the car example, the rest of the chemical energy is converted (mostly) into heat and sound energy. This is wasted energy — although you could always stick your dinner under the bonnet and warm it up on the drive home.
4) The less energy that is wasted, the more efficient the device is said to be.

Heat exchangers can be used to reduce the amount of energy wasted by a machine. They do this by pumping a cool fluid through the escaping heat and using it for other purposes. For example, the heat from a car's engine can be transferred to the air that's used to warm the passenger compartment.

A machine is a device which turns one type of energy into another. The efficiency of any device is defined as:

$$\text{Efficiency} = \frac{\text{Useful Energy out}}{\text{Total Energy in}}$$

You might not know the energy inputs and outputs of a machine, but you can still calculate the machine's efficiency as long as you know the power input and output:

$$\text{Efficiency} = \frac{\text{Useful Power out}}{\text{Total Power in}}$$

You can give efficiency as a decimal or you can multiply your answer by 100 to get a percentage, i.e. 0.75 or 75%.

As usual, a formula triangle will come handy for rearranging the formulas:

Useful Out
Efficiency × Total In

How to Use the Formula — Nothing to It

1) You find how much energy is supplied to a machine. (The Total Energy IN.)
2) You find how much useful energy the machine delivers. (The Useful Energy OUT.) An exam question might tell you this directly or tell you how much it wastes as heat/sound.
3) Either way, you get those two important numbers and then just divide the smaller one by the bigger one to get a value for efficiency somewhere between 0 and 1 (or 0 and 100%). Easy.
4) The other way they might ask it is to tell you the efficiency and the input energy and ask for the energy output — so you need to be able to swap the formula round.

Useful Energy Input Isn't Usually Equal to Total Energy Output

For any given example you can talk about the types of energy being input and output, but remember this:

No device is 100% efficient and the wasted energy is usually spread out as heat.

Electric heaters are the exception to this. They're usually 100% efficient because all the electricity is converted to "useful" heat. Ultimately, all energy ends up as heat energy. If you use an electric drill, it gives out various types of energy but they all quickly end up as heat.

Let there be light — and a bit of wasted heat...

The thing about loss of energy is it's always the same — it always disappears as heat and sound, and even the sound ends up as heat pretty quickly. So when they ask, "Why is the input energy more than the output energy?", the answer is always the same... Learn and enjoy.

Energy Transformation Diagrams

This is another opportunity for a MATHS question. Fantastic.
So best prepare yourself — here's what those energy transformation diagrams are all about...

The Thickness of the Arrow Represents the Amount of Energy

The idea of Sankey diagrams is to make it easy to see at a glance how much of the total energy in is being usefully employed compared with how much is being wasted.

The thicker the arrow, the more energy it represents — so you see a big thick arrow going in, then several smaller arrows going off it to show the different energy transformations taking place.

You can have either a little sketch or a properly detailed diagram where the width of each arrow is proportional to the number of joules it represents.

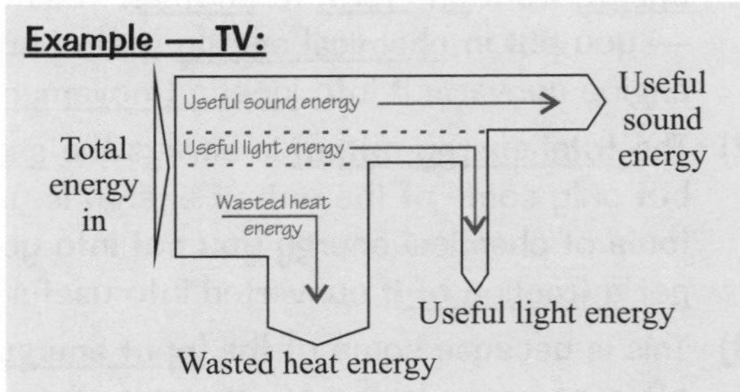

Example — Sankey Diagram for a Simple Motor:

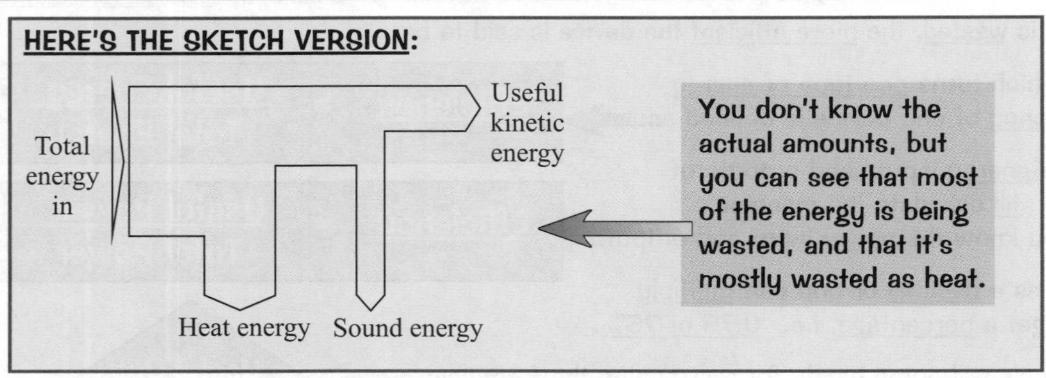

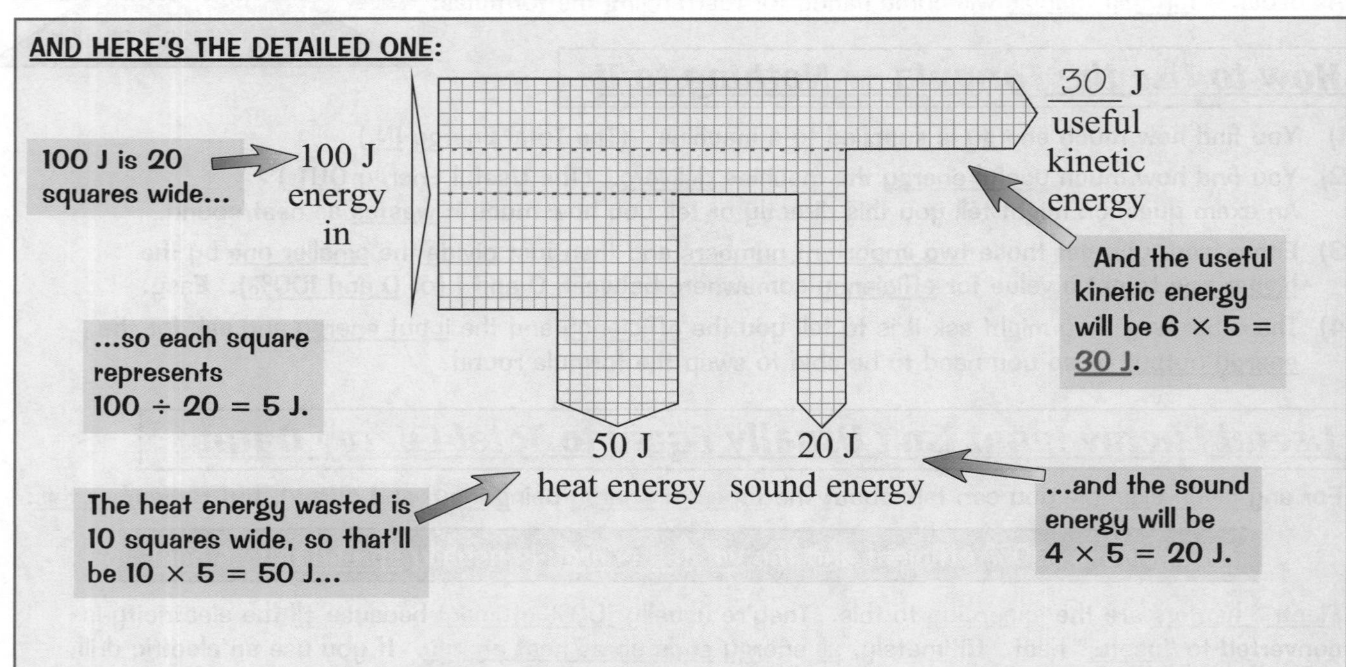

Skankey diagrams — to represent the smelliness of your socks...

If they ask you to draw your own Sankey diagram in the exam, and don't give you the figures, a sketch is all they'll expect. Just give a rough idea of where the energy goes. E.g. a filament lamp turns most of the input energy into heat, and only a tiny proportion goes to useful light energy.

Power Stations and Nuclear Energy

So you might know that electricity is a form of energy, but you're probably wondering how we get at it. No? Well read on anyway — that's what the next few pages are all about.

Non-Renewable Energy Resources Will Run Out One Day

The non-renewables are the three FOSSIL FUELS and NUCLEAR:

1) Coal
2) Oil
3) Natural gas
4) Nuclear fuels (uranium and plutonium)

a) They will all 'run out' one day.
b) They all do damage to the environment.
c) But they provide most of our energy.

Energy Sources can be Burned to Drive Turbines in Power Stations

Most of the electricity we use is generated from the four NON-RENEWABLE sources of energy (coal, oil, gas and nuclear) in big thermal power stations, which are all pretty much the same apart from the boiler.

1) The fossil fuel is burned to convert its stored chemical energy into heat (thermal) energy.
2) The heat energy is used to heat water (or air in some fossil-fuel power stations) to produce steam.
3) The steam turns a turbine, converting heat energy into kinetic energy.
4) The turbine is connected to a generator, which transfers kinetic energy into electrical energy.

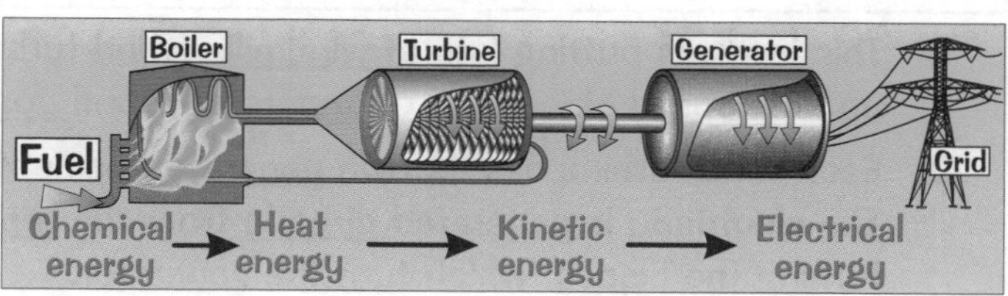

Nuclear Reactors are Just Fancy Boilers

A nuclear power station is mostly the same as the one above, but with nuclear fission of uranium or plutonium producing the heat to make steam to drive turbines, etc. The difference is in the boiler, as shown here:

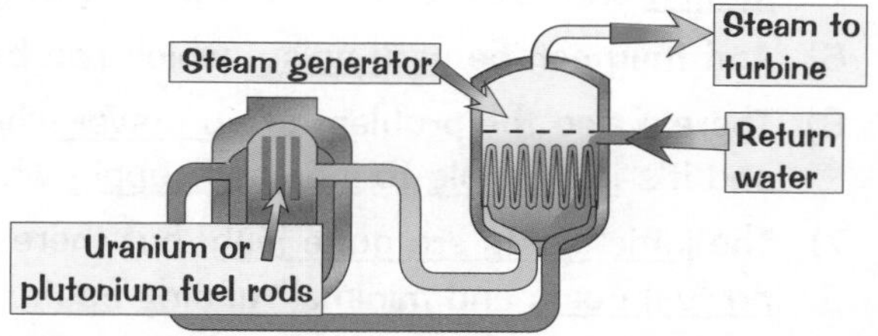

As with any type of energy source, there are advantages and disadvantages to using nuclear energy:

ADVANTAGES

1) Nuclear reactions release a lot more energy than chemical reactions.
2) A nuclear power station doesn't produce CO_2 when making electricity.
3) Nuclear fuel (e.g. uranium) is relatively cheap.

A gram of uranium releases over 10 000 times more energy than burning a gram of oil.

DISADVANTAGES

1) Nuclear power stations produce radioactive waste — this can be very dangerous and difficult to dispose of as it emits ionising radiation and stays radioactive for a long time (see p.78):
2) Radioactive waste can put people at risk through:
   - IRRADIATION — being exposed to radiation without coming into contact with the source. The damage to your body stops as soon as you leave the area where the radioactive waste is. You're only exposed for a short period of time, so will receive a lower dose of radiation.
   - CONTAMINATION — picking up some radioactive waste, e.g. by breathing it in, drinking contaminated water or getting it on your skin. You'll still be exposed to the radiation once you've left the area. Contamination is worse because it leaves people exposed to ionising damage for a long time, leading to more damage.
3) Nuclear power needs extra safety precautions — waste needs disposing of carefully (see p.78), the surrounding area needs to be tested for contamination of the soil and water, and workers need to be tested regularly to check they've not been exposed to too much radiation.
4) Nuclear power is supported by some people, but people who live close are often more scared of the risks.
5) Nuclear power stations take the longest time of all the power stations to start up. The overall cost of nuclear power is high due to the cost of the power plant and final decommissioning.

Renewable Energy Sources

Renewable energy sources, like wind and solar energy, will not run out. What's more, they do a lot less damage to the environment. They don't generate as much electricity as non-renewables though — if they did we'd all be using solar-powered toasters by now.

Renewable Energy Resources Will Never Run Out

The renewables are:

1) Wind
2) Waves
3) Tides
4) Hydroelectric
5) Solar
6) Geothermal
7) Food
8) Biofuels

a) These will never run out.
b) Most of them do damage the environment, but in less nasty ways than non-renewables.
c) The trouble is they don't provide much energy and some of them are unreliable because they depend on the weather.

Wind Power — Lots of Little Wind Turbines

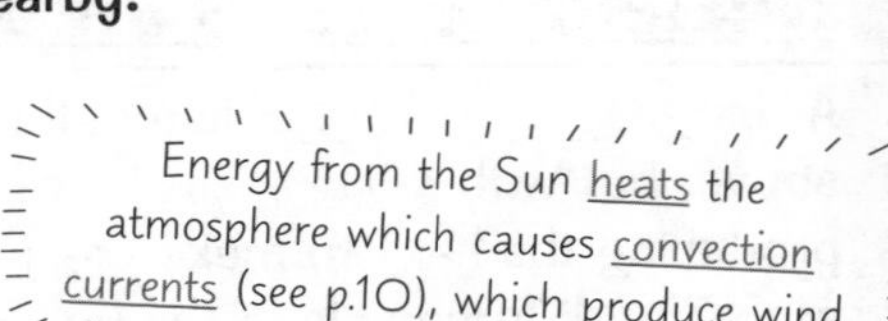

1) This involves putting lots of windmills (wind turbines) up in exposed places like on moors or round coasts.
2) Each wind turbine has its own generator inside it. The electricity is generated directly from the wind turning the blades, which turn the generator.
3) There's no pollution (except for a little bit when they're manufactured).
4) But they do spoil the view. You need about 1500 wind turbines to replace one coal-fired power station and 1500 of them cover a lot of ground — which would have a big effect on the scenery.
5) And they can be very noisy, which can be annoying for people living nearby.
6) There's also the problem of no power when the wind stops, and it's impossible to increase supply when there's extra demand.
7) The initial costs are quite high, but there are no fuel costs and minimal running costs.
8) There's no permanent damage to the landscape — if you remove the turbines, you remove the noise and the view returns to normal.

Energy from the Sun heats the atmosphere which causes convection currents (see p.10), which produce wind.

Solar Cells — Expensive but No Environmental Damage

(well, there may be a bit caused by making the cells)

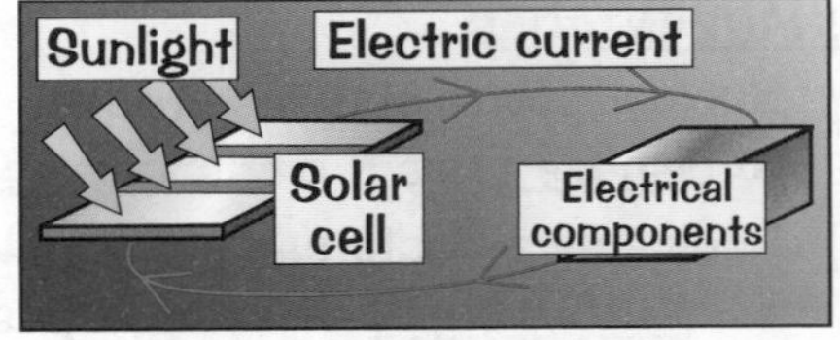

1) Solar cells generate electric currents directly from sunlight. Solar cells are often the best source of energy for calculators and watches which don't use much electricity.
2) Solar power is often used in remote places where there's not much choice (e.g. the Australian outback) and to power electric road signs and satellites.
3) There's no pollution. (Although they do use quite a lot of energy to manufacture in the first place.)
4) In sunny countries solar power is a very reliable source of energy — but only in the daytime. Solar power can still be cost-effective even in cloudy countries like Britain.
5) Initial costs are high but after that the energy is free and running costs almost nil.
6) Solar cells are usually used to generate electricity on a relatively small scale, e.g. powering individual homes.
7) It's often not practical or too expensive to connect them to the National Grid — the cost of connecting them to the National Grid can be enormous compared with the value of the electricity generated.

People love the idea of wind power — just not in their back yard...

Did you know you can now get rucksacks with built-in solar cells to charge up your mobile phone, MP3 player and digital camera while you're wandering around. Pretty cool, huh.

Renewable Energy Sources

Don't worry — I haven't forgotten about wave power and tidal power. It's easy to get confused between these two just because they're both to do with the seaside — but don't. They are completely different.

Wave Power — Lots of Little Wave-Powered Turbines

1) You need lots of small wave-powered turbines located around the coast.
2) As waves come in to the shore they provide an up and down motion which can be used to drive a generator.
3) There is no pollution. The main problems are spoiling the view and being a hazard to boats.
4) They are fairly unreliable, since waves tend to die out when the wind drops.
5) Initial costs are high, but there are no fuel costs and minimal running costs. Wave power is never likely to provide energy on a large scale, but it can be very useful on small islands.

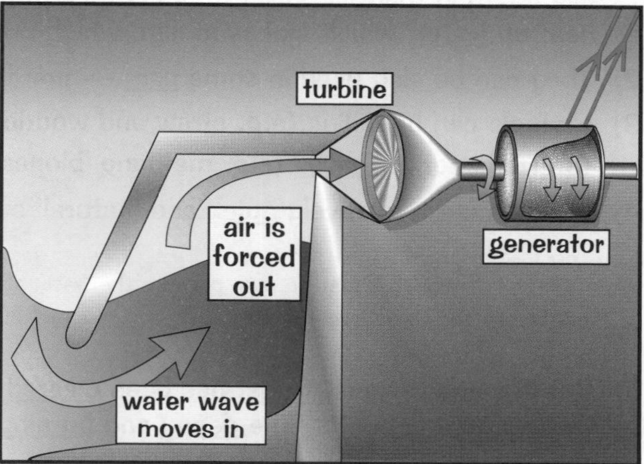

Tidal Barrages — Using the Sun and Moon's Gravity

1) Tidal barrages are big dams built across river estuaries, with turbines in them.
2) As the tide comes in it fills up the estuary to a height of several metres — it also drives the turbines. This water can then be allowed out through the turbines at a controlled speed.
3) The source of the energy is the gravity of the Sun and the Moon.
4) There is no pollution. The main problems are preventing free access by boats, spoiling the view and altering the habitat of the wildlife, e.g. wading birds, sea creatures and beasties who live in the sand.

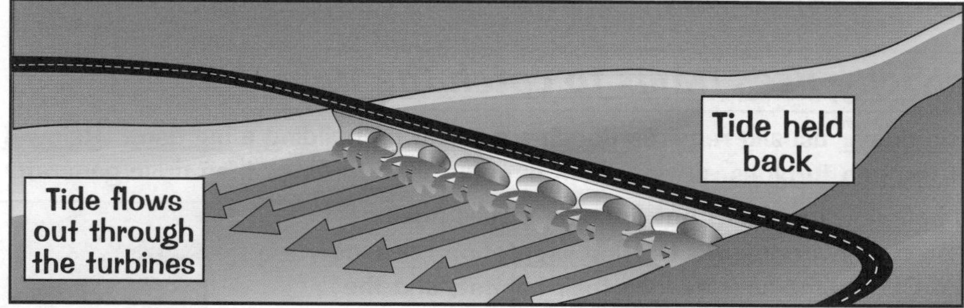

5) Tides are pretty reliable in the sense that they happen twice a day without fail, and always near to the predicted height. The only drawback is that the height of the tide is variable so lower (neap) tides will provide significantly less energy than the bigger 'spring' tides. They also don't work when the water level is the same either side of the barrage — this happens four times a day because of the tides. But tidal barrages are excellent for storing energy ready for periods of peak demand.
6) Initial costs are moderately high, but there are no fuel costs and minimal running costs. Even though it can only be used in some of the most suitable estuaries tidal power has the potential for generating a significant amount of energy.

Learn about Wave Power — and bid your cares goodbye...

I do hope you appreciate the big big differences between tidal power and wave power. They both involve salty seawater, sure — but there the similarities end. Lots of jolly details then, just waiting to be absorbed into your cavernous intracranial void. Smile and enjoy. And learn.

Renewable Energy Sources

Well, whaddaya know — there's more energy lurking about in piles of rubbish, rocks and rainwater.

Biofuels are Made from Plants and Waste

Biofuel is made from biomass — biological material from living organisms.

1) Biofuels are renewable energy resources. They're used to generate electricity in thermal power stations (see p.19) — they're burnt to heat up water, which makes steam, which drives a turbine.
2) They can be also used in some cars — just like fossil fuels.
3) Biofuels can be solids (e.g. straw and woodchips), liquids (e.g. ethanol) or gases (e.g. methane 'biogas' from sludge digesters).
4) Biofuels are a relatively quick and 'natural' source of energy and are supposedly carbon neutral:

> The plants that grew to produce the waste absorbed carbon dioxide from the atmosphere as they were growing. When the waste is burnt, this CO_2 is re-released into the atmosphere. So it has a neutral effect on atmospheric CO_2 levels.

5) But in some regions, forest has been cleared to make room to grow biofuels, resulting in species losing their natural habitats. The decay and burning of this vegetation also increases CO_2 and methane emissions.

Biofuel | Boiler | Turbine | Generator

Geothermal Energy — Heat from Underground

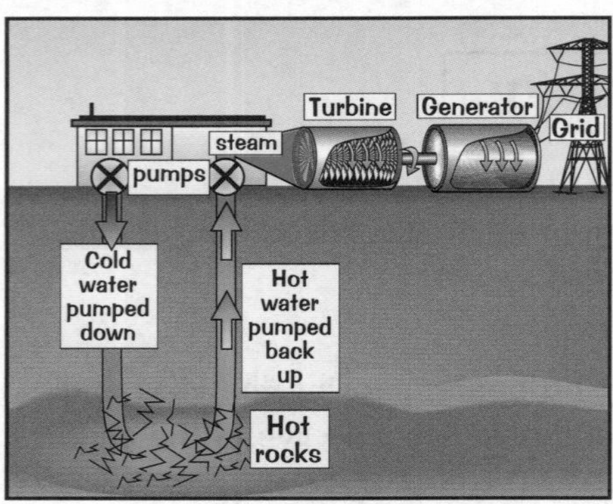

1) This is only possible where hot rocks lie quite near to the surface.
2) Geothermal energy is used to drive generators in thermal power stations. E.g. water is pumped in pipes down to hot rocks and returns as steam to drive a turbine.
3) This is actually brilliant free, renewable energy with no real environmental problems.
4) The main disadvantage is the cost of drilling down several km to the hot rocks.
5) Unfortunately there are very few places where this seems to be an economic option (for now).

Hydroelectricity uses Dams to catch the Rain

1) Hydroelectric power usually requires flooding a valley by building a big dam. Rainwater is caught and allowed out through turbines, driving them directly. The turbines then drive generators to make electricity.
2) There is no pollution (as such) but flooding a valley has a big impact on the environment. Rotting vegetation releases methane and CO_2, some species lose their habitat, and the reservoirs can also look very unsightly when they dry up. Location in remote valleys can reduce the human impact.
3) A big advantage is immediate response to increased demand, and it's fairly reliable except in times of drought.
4) Initial costs are high, but there's no fuel and minimal running costs.
5) You can draw a block diagram to show how hydroelectric power works:

dam | National Grid | water stored | turbines | generator

1) Water stored in reservoir above the turbines using a dam. → 2) Gravity causes the water to rush through the turbines. → 3) A generator converts the movement of the turbines (kinetic energy) into electricity.

Hydroelectric power — pretty dam good...

There's so much to learn on this page — I'll not keep you a minute longer. Get on and learn it...

Energy Sources and The Environment

Nothing worth having comes easily, and when generating electricity, you don't get a lot without affecting the environment. Luckily there are ways to get the most out of our resources — and this page explains how.

Pumped Storage Gives Extra Supply Just When It's Needed

1) Most large power stations have huge boilers which have to be kept running all night even though demand is very low. This means there's a surplus of electricity at night.
2) It's surprisingly difficult to find a way of storing this spare energy for later use.
3) Pumped storage is one of the best solutions.
4) In pumped storage, 'spare' night-time electricity is used to pump water up to a higher reservoir.
5) This can then be released quickly during periods of peak demand such as at teatime each evening, to supplement the steady delivery from the big power stations.
6) Remember, pumped storage uses the same idea as hydroelectric power, but it isn't a way of generating power — it's simply a way of storing energy which has already been generated.

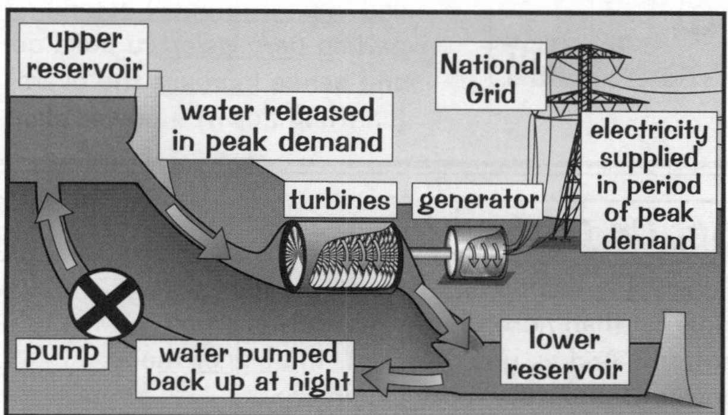

Non-Renewables are Linked to Environmental Problems

1) All three fossil fuels (coal, oil and gas) release CO_2 into the atmosphere when they're burned. For the same amount of energy produced, coal releases the most CO_2, followed by oil then gas. All this CO_2 adds to the greenhouse effect, and contributes to global warming (see p.38).
2) Burning coal and oil releases sulfur dioxide, which causes acid rain. Acid rain can be harmful to trees and soils and can have far-reaching effects in ecosystems.
3) Acid rain can be reduced by taking the sulfur out before the fuel is burned, or cleaning up the emissions.
4) Coal mining makes a mess of the landscape, especially "open-cast mining".
5) Oil spillages cause serious environmental problems, affecting mammals and birds that live in and around the sea. We try to avoid them, but they'll always happen.
6) Nuclear power is clean but the nuclear waste is very dangerous and difficult to dispose of (see p.78).

Carbon Capture can Reduce the Impact of Carbon Dioxide

1) Carbon capture and storage (CCS) is used to reduce the amount of CO_2 building up in the atmosphere and reduce the strength of the greenhouse effect.
2) CCS works by collecting the CO_2 from power stations before it is released into the atmosphere.
3) The captured CO_2 can then be pumped into empty gas fields and oil fields like those under the North Sea. It can be safely stored without it adding to the greenhouse effect.
4) CCS is a new technology that's developing quickly. New ways of storing CO_2 are being explored, including storing CO_2 dissolved in seawater at the bottom of the ocean and capturing CO_2 with algae, which can then be used to produce oil that can be used as a biofuel.

The power you're supplying — it's electrifying...

There's a lot to bear in mind with all the different energy sources and all the good things and nasty things associated with each of them. The next page is really handy for making comparisons between different energy sources.

Comparison of Energy Resources

Setting Up a Power Station

Because coal and oil are running out fast, many old coal- and oil-fired power stations are being taken out of use. Often they're being replaced by gas-fired power stations because they're quick to set up, there's still quite a lot of gas left and gas doesn't pollute as badly as coal and oil.

But gas is not the only option, as you really ought to know if you've been concentrating at all over the last few pages.

When looking at the options for a new power station, there are several factors to consider: How much it costs to set up and run, how long it takes to build, how much power it can generate, etc. Then there are also the trickier factors like damage to the environment and impact on local communities. And because these are often very contentious issues, getting permission to build certain types of power station can be a long-running process, and hence increase the overall set-up time. The time and cost of decommissioning (shutting down) a power plant can also be a crucial factor.

Set-Up Costs

Renewable resources often need bigger power stations than non-renewables for the same output. And as you'd expect, the bigger the power station, the more expensive.

Nuclear reactors and hydroelectric dams also need huge amounts of engineering to make them safe, which bumps up the cost.

Set-Up/Decommissioning Time

These are both affected by the size of the power station, the complexity of the engineering and also the planning issues (e.g. discussions over whether a nuclear power station should be built on a stretch of beautiful coastline can last years). Gas is one of the quickest to set up. Nuclear power stations take by far the longest (and cost the most) to decommission.

Reliability Issues

All the non-renewables are reliable energy providers (until they run out).

Many of the renewable sources depend on the weather, which means they're pretty unreliable here in the UK. The exceptions are tidal power and geothermal (which don't depend on weather).

Running/Fuel Costs

Renewables usually have the lowest running costs, because there's no actual fuel involved.

Environmental Issues

If there's a fuel involved, there'll be waste pollution and you'll be using up resources.

If it relies on the weather, it's often got to be in an exposed place where it sticks out like a sore thumb.

Atmospheric Pollution
Coal, Oil, Gas,
(+ others, though less so).

Visual Pollution
Coal, Oil, Gas, Nuclear, Tidal, Waves, Geothermal, Wind, Hydroelectric.

Other Problems
Nuclear (dangerous waste, explosions, contamination), Hydroelectric (dams bursting).

Using Up Resources
Coal, Oil, Gas, Nuclear.

Noise Pollution
Coal, Oil, Gas, Nuclear, Wind.

Disruption of Habitats
Hydroelectric, Tidal, Biofuels.

Disruption of Leisure Activities (e.g. boats)
Waves, Tidal.

Location Issues

This is fairly common sense — a power station has to be near to the stuff it runs on.

Solar — pretty much anywhere, though the sunnier the better

Gas — pretty much anywhere there's piped gas (most of the UK)

Hydroelectric — hilly, rainy places with floodable valleys, e.g. the Lake District, Scottish Highlands

Wind — exposed, windy places like moors and coasts or out at sea

Oil — near the coast (oil transported by sea)

Waves — on the coast

Coal — near coal mines, e.g. Yorkshire, Wales

Nuclear — away from people (in case of disaster), near water (for cooling)

Tidal — big river estuaries where a dam can be built

Geothermal — fairly limited, only in places where hot rocks are near the Earth's surface

Of course — the biggest problem is we need too much electricity...

It would be lovely if we could get rid of all the nasty polluting power stations and replace them with clean, green fuel, just like that... but it's not quite that simple. Renewable energy has its own problems too, and probably isn't enough to power the whole country without having a wind farm in everyone's back yard.

Electricity and The National Grid

The National Grid is the network of pylons and cables that covers the whole of Britain, getting electricity to homes everywhere. Whoever you pay for your electricity, it's the National Grid that gets it to you.

Electricity is Distributed via the National Grid...

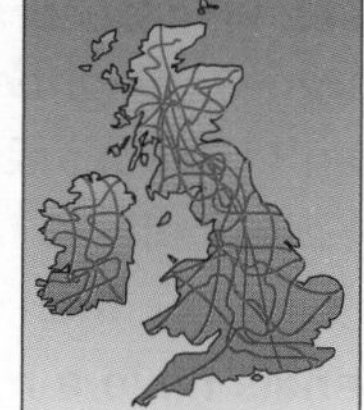

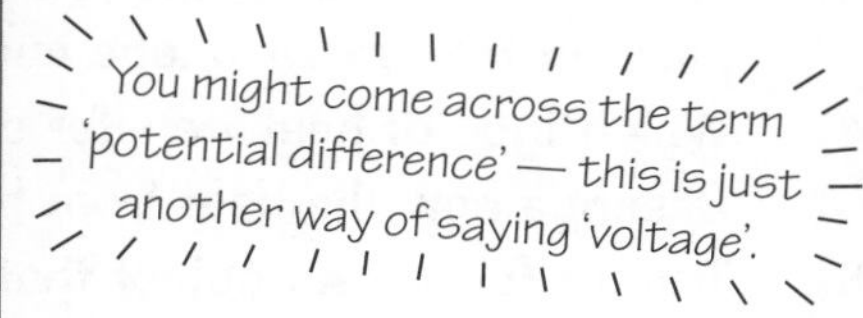

1) The National Grid takes electrical energy from power stations to where it's needed in homes and industry.
2) It enables power to be generated anywhere on the grid, and then be supplied anywhere else on the grid.
3) To transmit the huge amount of power needed, you need either a high voltage or a high current.
4) The problem with a high current is that you lose loads of energy through heat in the cables.
5) It's much cheaper to boost the voltage up really high (to 400 000 V) and keep the current very low (see page 86).

...With a Little Help from Pylons and Transformers

1) To get the voltage to 400 000 V to transmit power requires transformers as well as big pylons with huge insulators — but it's still cheaper.
2) The transformers have to step the voltage up at one end, for efficient transmission, and then bring it back down to safe, usable levels at the other end.
3) The voltage is increased ('stepped up') using a step-up transformer.
4) It's then reduced again ('stepped down') at the consumer end using a step-down transformer.

See pages 85-86 for more on transformers.

There are Different Ways to Transmit Electricity

1) Electrical energy can be moved around by cables buried in the ground, as well as in overhead power lines.
2) Each of these different options has its pros and cons:

	Setup cost	Maintenance	Faults	How it looks	Affected by weather	Reliability	How easy to set up	Disturbance to land
Overhead Cables	lower	lots needed	easy to access	ugly	yes	less reliable	easy	minimal
Underground Cables	higher	minimal	hard to access	hidden	no	more reliable	hard	lots

Supply and Demand

1) The National Grid needs to generate and direct all the energy that the country needs — our energy demands keep on increasing too.
2) In order to meet these demands in the future, the energy supplied to the National Grid will need to increase, or the energy demands of consumers will need to decrease.
3) In the future, supply can be increased by opening more power plants or increasing their power output (or by doing both).
4) Demand can be reduced by consumers using more energy-efficient appliances, and being more careful not to waste energy in the home (e.g. turning off the lights or running washing machines at cooler temperatures).

400 000 V — you wouldn't want to fly your kite into that...

If you had your own solar panel or wind generator, you could sell back any surplus electricity to the National Grid. So if you don't use much electricity, but you generate a lot of it, you can actually make money instead of spending it. Nice trick if you can do it. Shame solar panels cost a fortune...

Revision Summary for Section One

It's all very well reading the pages and looking at the diagrams — but you won't have a hope of remembering it for your exam if you don't understand it. Have a go at these questions to see how much has gone in so far.

1) Explain the difference between heat and temperature. What units are they each measured in?
2) Describe the arrangement and movement of the particles in a) solids b) liquids c) gases
3) What is the name of the process where vibrating particles pass on their extra kinetic energy to neighbouring particles?
4) Which type of heat transfer can't take place in solids — convection or conduction?
5) Describe how the heat from heater coils is transferred throughout the water in a kettle.
6) True or false? An object that's cooler than its surroundings emits more radiation than it absorbs.
7) Explain why solar hot water panels have a matt black surface.
8) Describe an experiment to demonstrate how the colour or texture of a material affects the amount of heat energy that it absorbs.
9) What happens to the particles of a gas as it turns to a liquid?
10) What is the name given to the process where a gas turns to a liquid?
11) The two designs of car engine shown are made from the same material. Which engine will transfer heat quicker? Explain why.

Engine A Engine B

12) Describe two features of a vacuum flask that make it good at keeping hot liquids hot.

13) Do animals that live in hot climates tend to have large or small ears? Give one reason why this might be an advantage in a hot climate.

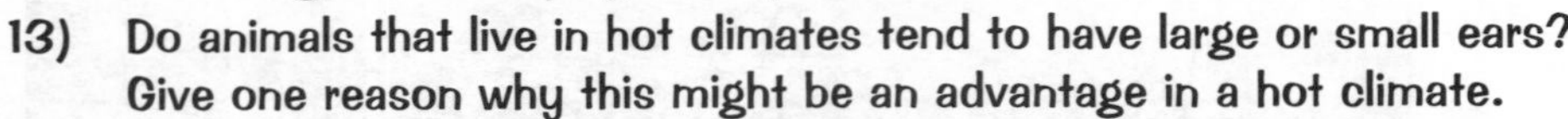

14)* If it costs £4000 to double glaze your house and the double glazing saves you £100 on energy bills every year, calculate the payback time for double glazing.

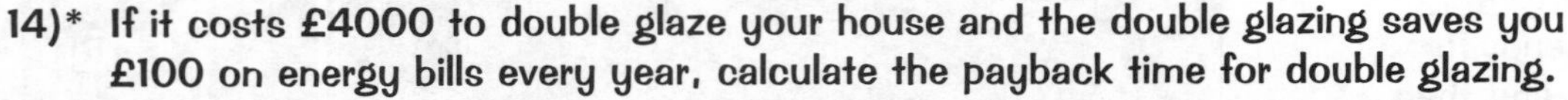

15) Name five ways of improving energy efficiency in the home. Explain how each improvement reduces the amount of heat lost from a house.
16) What can you tell from a material's U-value?
17) Would you expect copper or cotton wool to have a higher U-value?
18) What property of a material tells you how much energy it can store?
19)* An ornament has a mass of 0.5 kg. The ornament is made from a material that has a specific heat capacity of 1000 J/kg°C. How much energy does it take to heat the ornament from 20 °C to 200 °C?
20) Do heaters use materials that have a high or low heat capacity?
21) Name nine types of energy and give an example of each.
22) State the principle of the conservation of energy.
23) List the energy transformations that occur in a battery-powered toy car.
24) What is the useful type of energy delivered by a motor? In what form is energy wasted?
25)* What is the efficiency of a motor that converts 100 J of electrical energy into 70 J of useful kinetic energy?
26)* The following Sankey diagram shows how energy is converted in a catapult.

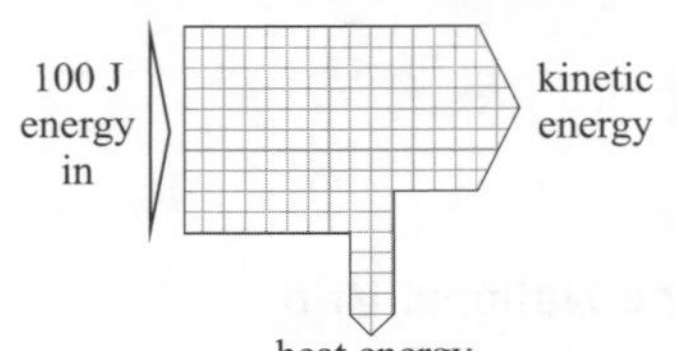

a) How much energy is converted into kinetic energy?
b) How much energy is wasted?
c) What is the efficiency of the catapult?

27) Describe the main features of a nuclear power station.
28) List three advantages and three disadvantages of using nuclear power.
29) Explain the difference between irradiation and contamination.
30) Name two sources of energy that power turbines directly without the use of steam.
31) Give one advantage and one disadvantage for using the following types of power: a) wave, b) tidal, c) wind, d) solar, e) biofuels, f) geothermal.
32) List four disadvantages of using fossil fuels as an energy source. Why do we still need to use them?
33) What is the National Grid?
34) Explain why a very high voltage is used to transmit electricity in the National Grid.

*Answers on page 140.

Wave Basics

Waves transfer energy from one place to another without transferring any matter (stuff).

Waves Have Amplitude, Wavelength and Frequency

1) The amplitude is the displacement from the rest position to the crest (NOT from a trough to a crest).
2) The wavelength is the length of a full cycle of the wave, e.g. from crest to crest.
3) Frequency is the number of complete waves passing a certain point per second OR the number of waves produced by a source each second. Frequency is measured in hertz (Hz). 1 Hz is 1 wave per second.
4) The speed is, well, how fast it goes.
5) You can work out the distance a wave has travelled, how long it took or the speed it was going by using the formula:

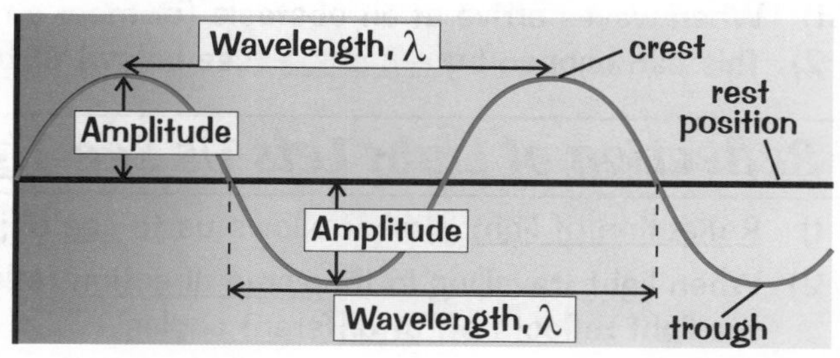

distance = speed × time
(m) (m/s) (s)

Transverse Waves Have Sideways Vibrations

Most waves are transverse:

1) Light and all other electromagnetic (EM) waves (p.32).
2) A slinky spring wiggled up and down.
3) Waves on strings.
4) S-waves (see p.113).

In TRANSVERSE waves the vibrations are PERPENDICULAR (at 90°) to the DIRECTION OF ENERGY TRANSFER of the wave.

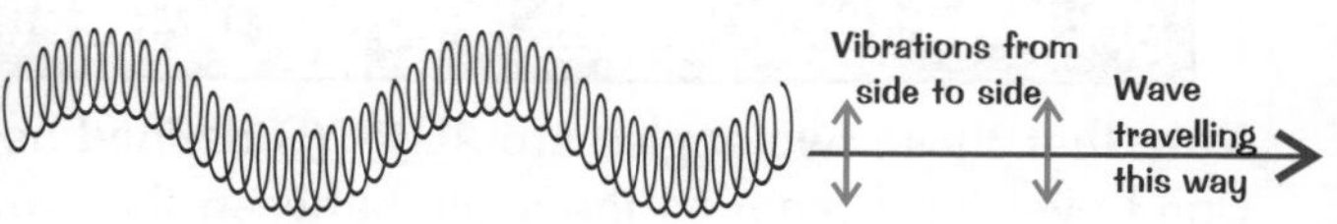

Longitudinal Waves Have Vibrations Along the Same Line

Examples of longitudinal waves are:

1) Sound waves and ultrasound.
2) P-waves (see p.113).
3) A slinky spring when you push the end.

Water waves, P-waves, S-waves and waves in springs and ropes are all examples of mechanical waves.

In LONGITUDINAL waves the vibrations are PARALLEL to the DIRECTION OF ENERGY TRANSFER of the wave.

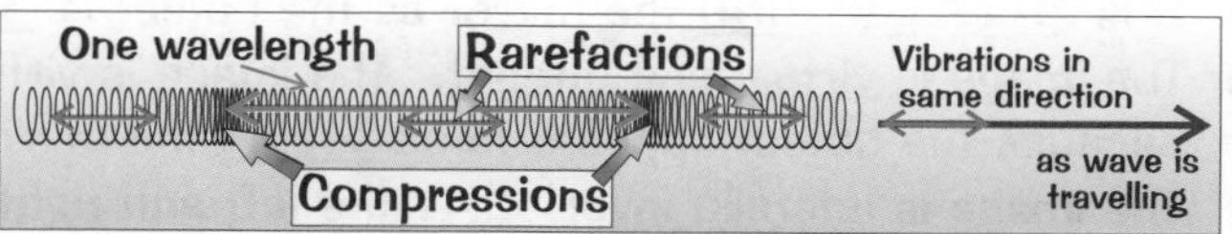

Wave Speed = Frequency × Wavelength

The equation below applies to all waves. You need to practise using it.

Speed = Frequency × Wavelength
(m/s) (Hz) (m)

OR $v = f \times \lambda$

Speed (v is for velocity)
Frequency
Wavelength (that's the Greek letter 'lambda')

$\frac{v}{f \times \lambda}$

EXAMPLE: A radio wave has a frequency of 92.2×10^6 Hz. Find its wavelength. (The speed of all EM waves is 3×10^8 m/s.)

ANSWER: You're trying to find λ using f and v, so you've got to rearrange the equation. So $\lambda = v \div f = 3 \times 10^8 \div 9.22 \times 10^7 = 3.25$ m.

The speed of a wave is usually independent of the frequency or amplitude of the wave.

Waves — dig the vibes, man...

Get that diagram at the top of the page learned, then get that $v = f \times \lambda$ business imprinted on your brain. When you've done that, try this question: A sound wave travelling in a solid has a frequency of 1.9×10^4 Hz and a wavelength of 12.5 cm. Find its speed.*

* Answer on p.140

Wave Properties

If you're anything like me, you'll have spent hours gazing into a mirror in wonder. Here's why...

All Waves Can be Reflected, Refracted and Diffracted

1) When waves arrive at an obstacle (or meet a new material), their direction of travel can be changed.
2) This can happen by reflection (see below) or by refraction or diffraction (see p.29).

Reflection of Light Lets Us See Things

1) Reflection of light is what allows us to see objects. Light bounces off them into our eyes.
2) When light travelling in the same direction reflects from an uneven surface such as a piece of paper, the light reflects off at different angles.
3) When light travelling in the same direction reflects from an even surface (smooth and shiny like a mirror) then it's all reflected at the same angle and you get a clear reflection.

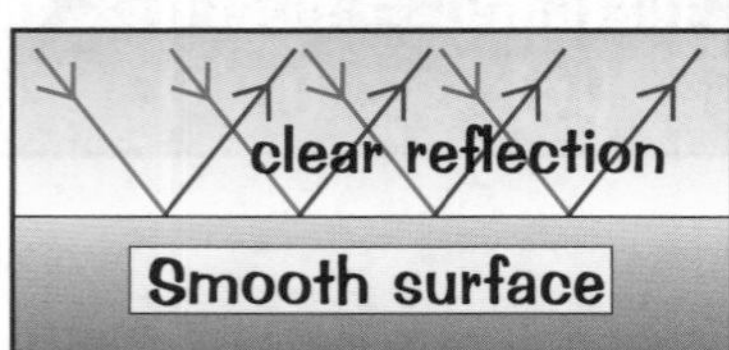

The normal is an imaginary line that's perpendicular (at right angles) to the surface at the point of incidence (where the light hits the surface).

Shiny side of mirror
Dull side of mirror
Reflected light ray
angle of reflection, r
Normal
angle of incidence, i
Incident light ray

4) The LAW OF REFLECTION applies to every reflected ray:

Angle of INCIDENCE = Angle of REFLECTION

Note that these two angles are ALWAYS defined between the ray itself and the NORMAL, dotted above. Don't ever label them as the angle between the ray and the surface. Definitely uncool.

Draw a Ray Diagram for an Image in a Plane Mirror

I said ANGLE

The diagrams below show how an image is formed in a PLANE MIRROR. Here are the important points:

1) The image is the same size as the object.
2) It is AS FAR BEHIND the mirror as the object is in front.
3) The image is virtual and upright. The image is virtual because the object appears to be behind the mirror.
4) The image is laterally inverted — the left and right sides are swapped, i.e. the object's left side becomes its right side in the image.

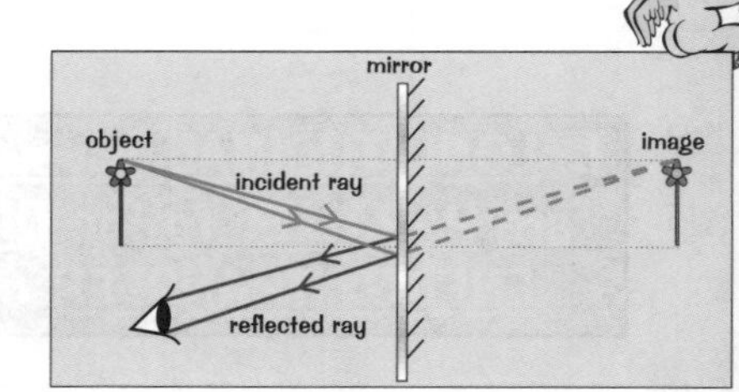

1) First off, draw the virtual image. Don't try to draw the rays first. Follow the rules in the above box — the image is the same size, and it's as far behind the mirror as the object is in front.
2) Next, draw a reflected ray going from the top of the virtual image to the top of the eye. Draw a bold line for the part of the ray between the mirror and eye, and a dotted line for the part of the ray between the mirror and virtual image.
3) Now draw the incident ray going from the top of the object to the mirror. The incident and reflected rays follow the law of reflection — but you don't actually have to measure any angles. Just draw the ray from the object to the point where the reflected ray meets the mirror.
4) Now you have an incident ray and reflected ray for the top of the image. Do steps 2 and 3 again for the bottom of the eye — a reflected ray going from the image to the bottom of the eye, then an incident ray from the object to the mirror.

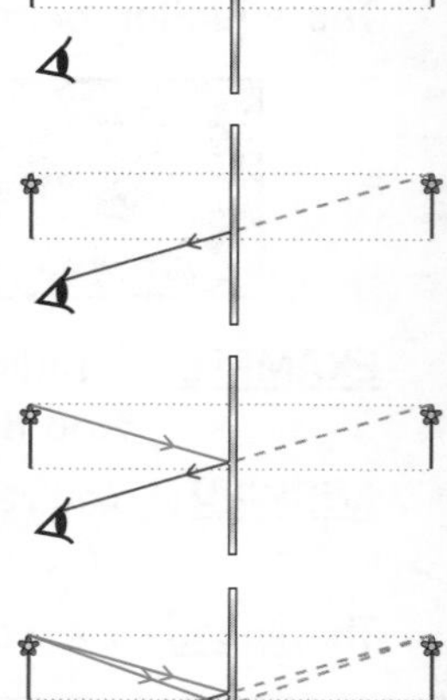

Plane mirrors — what pilots use to look behind them...

Make sure you can draw clear ray diagrams — they're a favourite with examiners across the Universe.

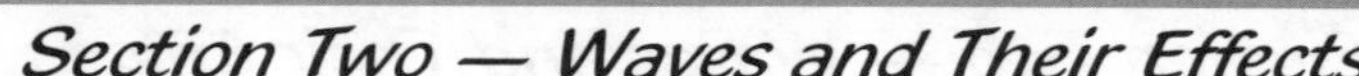

Diffraction and Refraction

If you thought reflection was good, you'll just love diffraction and refraction — they're awesome. If you didn't find reflection interesting then I'm afraid it's tough luck — you'll have to read about them anyway. Sorry.

Diffraction — Waves Spreading Out

1) All waves spread out ('diffract') at the edges when they pass through a gap or pass an object.
2) The amount of diffraction depends on the size of the gap relative to the wavelength of the wave. The narrower the gap, or the longer the wavelength, the more the wave spreads out.
3) A narrow gap is one about the same size as the wavelength of the wave. So whether a gap counts as narrow or not depends on the wave.

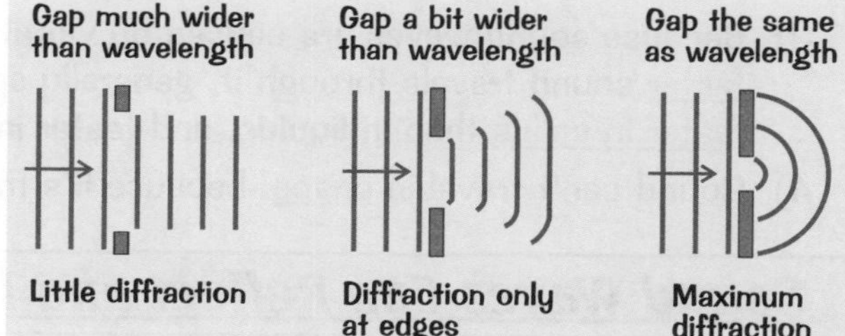

4) Light has a very small wavelength (about 0.0005 mm), so it can be diffracted but it needs a really small gap.
5) This means you can hear someone through an open door even if you can't see them, because the size of the gap and the wavelength of sound are roughly equal, causing the sound wave to diffract and fill the room...
6) ...But you can't see them unless you're directly facing the door because the gap is about a million times bigger than the wavelength of light, so it won't diffract enough.
7) If a gap is about the same size as the wavelength of a light, you can get a diffraction pattern of light and dark fringes, as shown here.

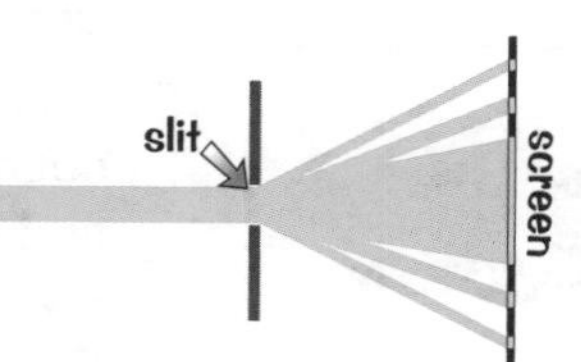

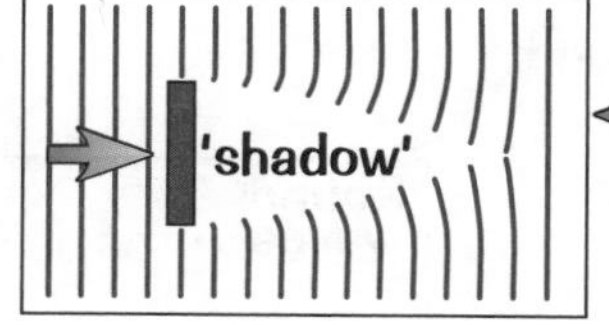

8) You get diffraction around the edges of obstacles too. The shadow is where the wave is blocked. The wider the obstacle compared to the wavelength, the less diffraction it causes, so the longer the shadow.

See p.110 for more on diffraction.

Refraction — Changing the Speed of a Wave Can Change its Direction

1) Waves travel at different speeds in substances which have different densities. So when a wave crosses a boundary between two substances (from glass to air, say) it changes speed:

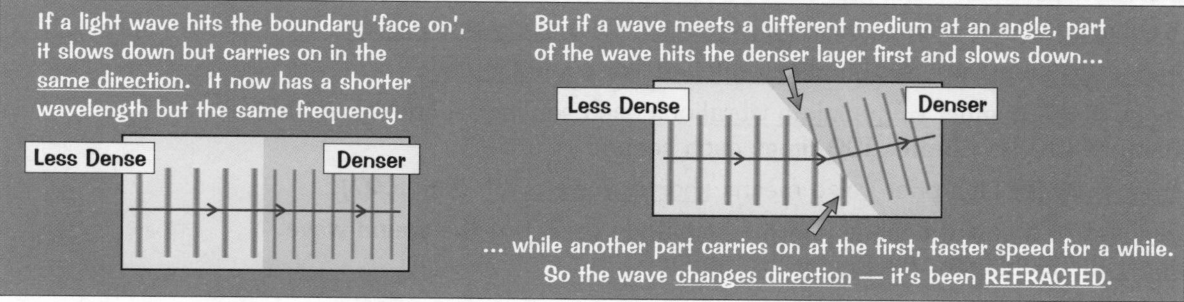

2) E.g. when light passes from air into the glass of a window pane (a denser medium), it slows down — causing the light to refract towards the normal. When the light reaches the 'glass to air' boundary on the other side of the window, it speeds up and refracts away from the normal.

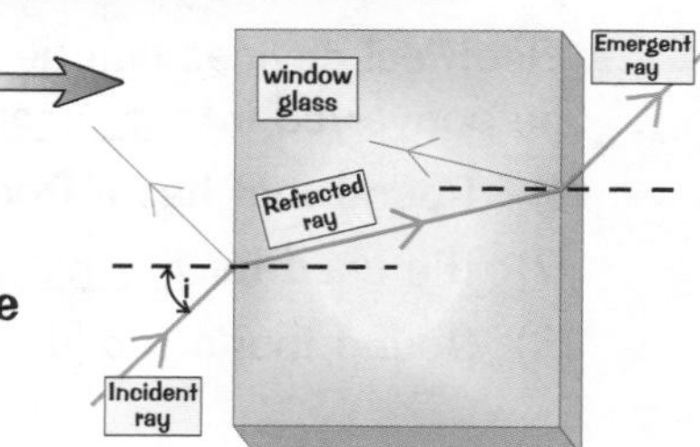

3) Waves are only refracted if they meet a new medium at an angle. If they're travelling along the normal (i.e. the angle of incidence is zero) they will change speed, but are NOT refracted — they don't change direction.

Lights, camera, refraction...

Remember that all waves can be diffracted. It doesn't matter what type of wave it is — sound, light, water... The key point to remember about refraction is that the wave has to meet a boundary at an angle.

Sound Waves

We hear sounds when vibrations reach our eardrums. Read on to find out how sound waves work.

Sound Travels as a Wave

1) Sound waves are caused by vibrating objects. These mechanical vibrations are passed through the surrounding medium as a series of compressions. They're a type of longitudinal wave (see p.27).

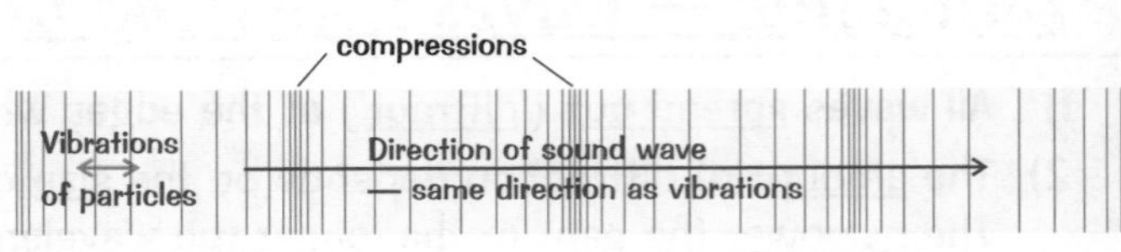

2) Sometimes the sound will eventually travel through someone's inner ear and reach their eardrum, at which point the person might hear it.
3) Because sound waves are caused by vibrating particles, the denser the medium, the faster sound travels through it, generally speaking anyway. Sound generally travels faster in solids than in liquids, and faster in liquids than in gases.
4) Sound can't travel in space, because it's mostly a vacuum (there are no particles).

Sound Waves Can Reflect and Refract

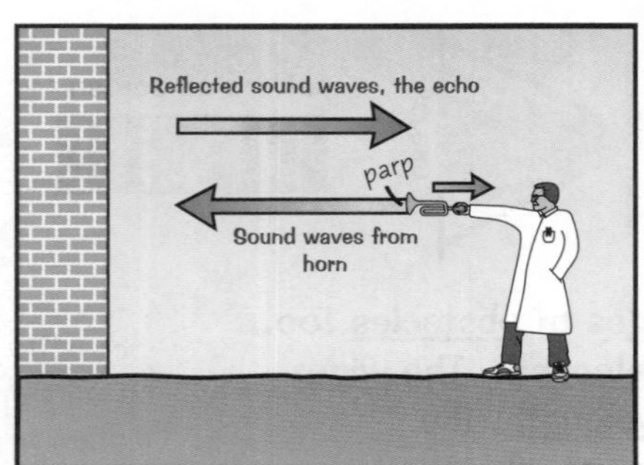

1) Sound waves will be reflected by hard flat surfaces.
2) This is very noticeable in an empty room. A big empty room sounds completely different once you've put carpet, curtains and a bit of furniture in it. That's because these things absorb the sound quickly and stop it echoing around the room. Echoes are just reflected sound waves.
3) You hear a delay between the original sound and the echo because the echoed sound waves have to travel further, and so take longer to reach your ears.

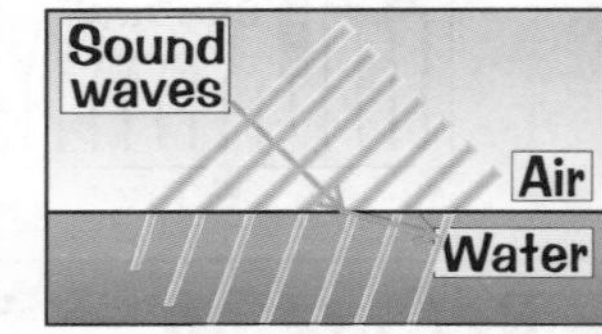

4) Sound waves will also refract (see page 29) as they enter different media. As they enter denser material, they speed up. (However, since sound waves are always spreading out so much, the change in direction is hard to spot under normal circumstances.)

The Higher the Frequency, the Higher the Pitch

1) High frequency sound waves sound high-pitched like a squeaking mouse.
2) Low frequency sound waves sound low-pitched like a mooing cow.
3) Frequency is the number of complete vibrations each second — so a wave that has a frequency of 100 Hz vibrates 100 times each second.
4) High frequency (or high pitch) also means shorter wavelength (see p.27).
5) The loudness of a sound depends on the amplitude (p.27) of the sound wave. The bigger the amplitude, the louder the sound.

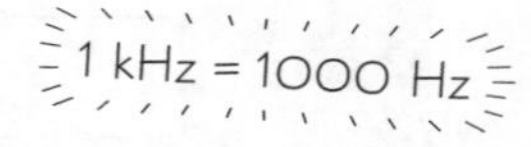

Electrical devices can be made which produce electrical oscillations of any frequency. These can easily be converted into mechanical vibrations to produce sound waves beyond the range of human hearing:

1) Sound that has a frequency above the range of human hearing (about 20 kHz) is called ultrasound.
2) It's used for pre-natal scanning, and sonar.
3) Sound that's too low for humans to hear (below 20 Hz) is called infrasound.

The room always feels big and empty whenever I tell a joke... (It must be the carpets.)

So, sound is the vibrations of particles, and travels as a longitudinal wave. When sound reaches your ear, the air particles in your ear canal vibrate against your eardrum, which vibrates against three tiny bones. These set inner-ear fluid moving, which moves thousands of delicate cells, which send signals to your brain.

Analogue and Digital Signals

Sound and images can be sent as analogue or digital signals, but digital technology is gradually taking over.

Information is Converted into Signals

Information is being transmitted everywhere all the time.

1) Whatever kind of information you're sending (text, sound, pictures...) it's converted into electrical signals before it's transmitted.
2) It's then sent long distances down telephone lines or...
3) ...superimposed (mixed) onto 'carrier' EM waves (see next page).
4) It's then sent out as either analogue or digital signals.

Analogue Signals Vary but Digital's Either On or Off

1) The amplitude or frequency of an analogue signal varies continuously. An analogue signal can take any value in a particular range.
2) Digital signals can only take one of a small number of discrete values (usually two), e.g. 0 or 1, on or off, true or false.
3) The information is carried by switching the EM carrier wave on or off.
4) This creates pulses — short bursts of waves, e.g. where 0 = off (no pulse) and 1 = on (pulse).
5) A digital receiver will decode these pulses to get a copy of the original signal.

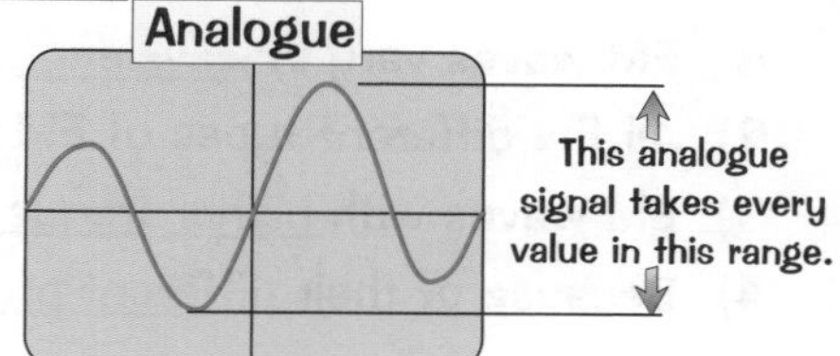

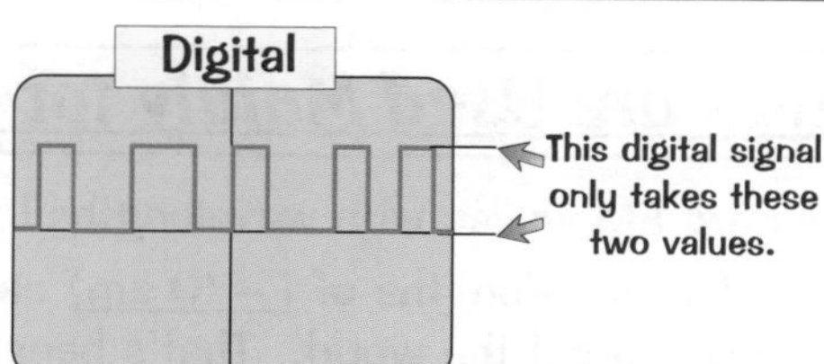

Signals Have to be Amplified

Both digital and analogue signals weaken as they travel, so they may need to be amplified along their route. They also pick up interference or noise from electrical disturbances or other signals.

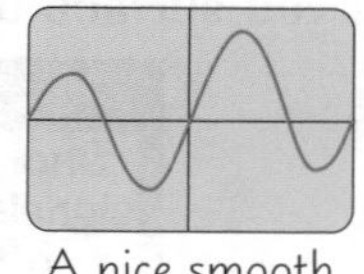

A nice smooth analogue signal

The same signal with noise

Digital Signals are Far Better Quality

1) Noise is less of a problem with digital signals than with analogue. If you receive a 'noisy' digital signal, it's pretty obvious what it's supposed to be. So it's easy to 'clean up' the signal — the noise doesn't get amplified.
2) But if you receive a noisy analogue signal, it's difficult to know what the original signal would have looked like. And if you amplify a noisy analogue signal, you amplify the noise as well.

This noisy digital signal...

...is obviously supposed to be this.

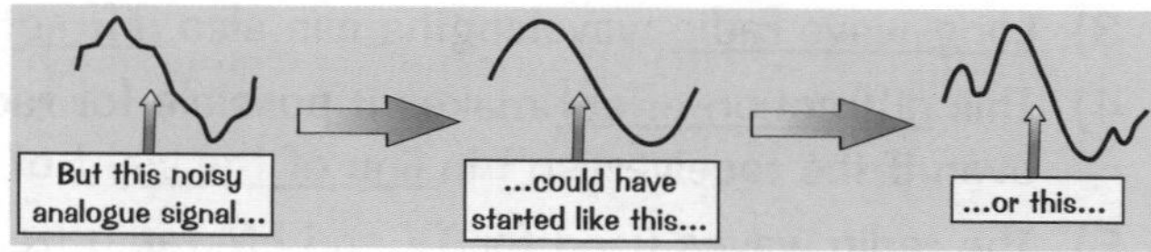

3) This is why digital signals are much higher quality — the information received is the same as the original.
4) Another advantage of digital technology is you can transmit several signals at once using just one cable or EM wave — this is called multiplexing — so you can send more information (in a given time) than using analogue signals.
5) Also, digital signals are easy to process using computers, since computers are digital devices.
6) The amount of information used to store a digital image or sound is measured in bytes. Images and sounds will be of higher quality when the amount of information stored is higher.

I've got loads of digital stuff — watch, radio, fingers...

Digital signals are great — unless you live in a part of the country which currently has poor reception of digital broadcasts, in which case you get no benefit at all. This is because if you don't get spot-on reception of digital signals in your area, you won't get a grainy picture (as with analogue signals) — you'll get nothing at all.

EM Waves and Communication

Types of electromagnetic (EM) wave have a lot in common with one another, but their differences make them useful to us in different ways. These pages are packed with loads of dead useful info, so pay attention...

There's a Continuous Spectrum of EM Waves

EM waves with different wavelengths (or frequencies) have different properties. We group them into seven basic types, but the different regions actually merge to form a continuous spectrum.

They're shown below with increasing frequency and energy (decreasing wavelength) from left to right.

	RADIO WAVES	MICRO WAVES	INFRA RED	VISIBLE LIGHT	ULTRA VIOLET	X-RAYS	GAMMA RAYS
wavelength →	1 m – 10^4 m	10^{-2} m (1 cm)	10^{-5} m (0.01 mm)	10^{-7} m	10^{-8} m	10^{-10} m	10^{-15} m

1) EM waves vary in wavelength from around 10^{-15} m to more than 10^4 m.
2) All the different types of EM wave travel at the same speed (3×10^8 m/s) in a vacuum (e.g. space).
3) EM waves with higher frequencies have shorter wavelengths.
4) Because of their different properties, different EM waves are used for different purposes.

Radio Waves are Used Mainly for Communication

1) Radio waves are EM waves with wavelengths longer than about 10 cm.
2) Long-wave radio (wavelengths of 1 – 10 km) can be transmitted from London, say, and received halfway round the world. That's because long wavelengths diffract (bend) (see p.29) around the curved surface of the Earth.

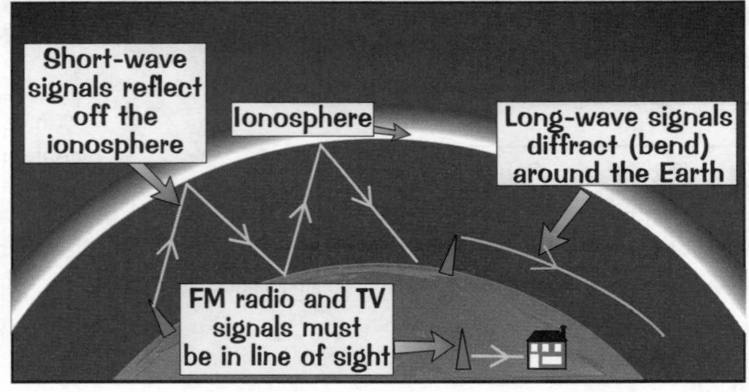

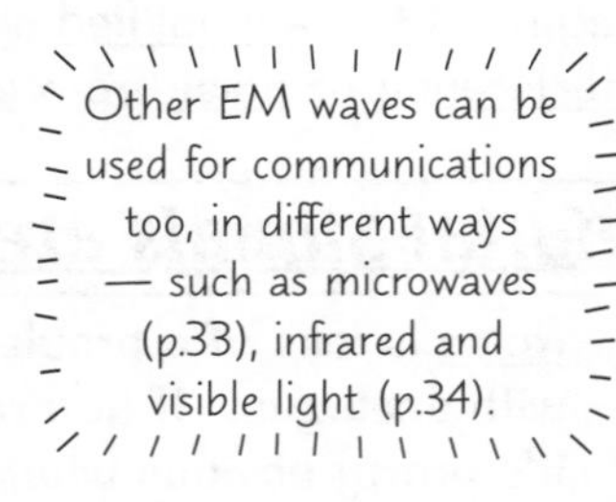

3) Long-wave radio wavelengths can also diffract around hills, into tunnels and all sorts.
4) This diffraction effect makes it possible for radio signals to be received even if the receiver isn't in line of the sight of the transmitter.
5) The radio waves used for TV and FM radio transmissions have very short wavelengths (10 cm – 10 m). To get reception, you must be in direct sight of the transmitter — the signal doesn't bend around hills or travel far through buildings.
6) Short-wave radio signals (wavelengths of about 10 m – 100 m) can, like long-wave, be received at long distances from the transmitter. That's because they are reflected (p.28) off the ionosphere — an electrically charged layer in the Earth's upper atmosphere.
7) Medium-wave signals (well, the shorter ones) can also reflect from the ionosphere, depending on atmospheric conditions and the time of day.

Size matters — and my wave's longer than yours...

You might have to name and order the different types of EM waves in terms of their energy, frequency and wavelength. To remember the order of increasing frequency and energy, I use the mnemonic Rock Music Is Very Useful for eXperiments with Goats. It sounds stupid but it does work — why not make up your own...

Microwaves

No phones, no dinner — what would we do without microwaves?

Microwaves are Used for Satellite Communication and Mobile Phones

1) Communication to and from satellites (including satellite TV signals and satellite phones) uses microwaves. But you need to use microwaves which can pass easily through the Earth's watery atmosphere.
2) For satellite TV, the signal from a transmitter is transmitted into space...
3) ... where it's picked up by the satellite receiver dish orbiting thousands of kilometres above the Earth. The satellite transmits the signal back to Earth in a different direction...
4) ... where it's received by a satellite dish on the ground. There is a slight time delay between the signal being sent and received, e.g. from the UK to Australia, because of the long distance the signal has to travel.
5) Mobile phone signals also travel from your phone to the nearest transmitter as microwaves.

Microwave Ovens Use a Different Microwave Wavelength from Satellites

1) In communications, the microwaves used need to pass through the Earth's watery atmosphere.
2) In microwave ovens, the microwaves need to be absorbed by water molecules in food to be able to heat it up — so they use a different wavelength to those used in satellite communications.
3) The microwaves penetrate up to a few centimetres into the food before being absorbed by water molecules. The energy from the absorbed microwaves causes the food to heat up. The heat energy is then conducted or convected to other parts of the food.

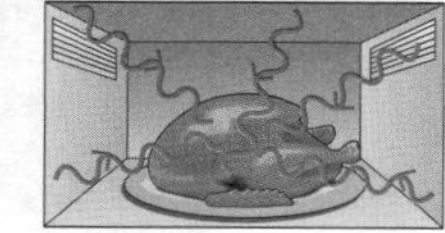

Microwave radiation would heat up the water in your body's cells if you were exposed to it. Microwave ovens have metal cases and screens over their glass doors which reflect and absorb the microwaves, stopping them getting out.

Some People Say There are Health Risks with Using Microwaves

1) When you make a call on your mobile, the phone emits microwave radiation. If some of this radiation were to be absorbed by your body, and cause heating of your body tissues (which all contain water), cells could be burned or killed.
2) Some people think that the microwaves emitted into your body from using a mobile phone or living near a mobile phone mast could damage your health.
3) There's no conclusive proof either way yet though. Lots of studies have been published, which has allowed the results to be checked, but so far they have given conflicting evidence.

Microwaves — for when you're only slightly sad to say goodbye...

In a microwave, the whole idea is for water molecules to absorb the microwave energy. But that's not ideal for communications — if all the energy was absorbed by water molecules, the signals would never get through clouds.

Infrared and Visible Light

Infrared radiation (or IR) may sound space-age but it's actually as common as beans on toast.

Infrared Waves are Used for Remote Controls and Optical Fibres

1) Infrared waves are used in lots of wireless remote controllers.
2) Remote controls work by emitting different patterns of infrared waves to send different commands to an appliance, e.g. a TV.
3) Optical fibres (e.g. those used in phone lines) can carry data over long distances very quickly.
4) They use both infrared waves and visible light.
5) The signal is carried as pulses of light or infrared radiation and is reflected off the sides of a very narrow core from one end of the fibre to the other.

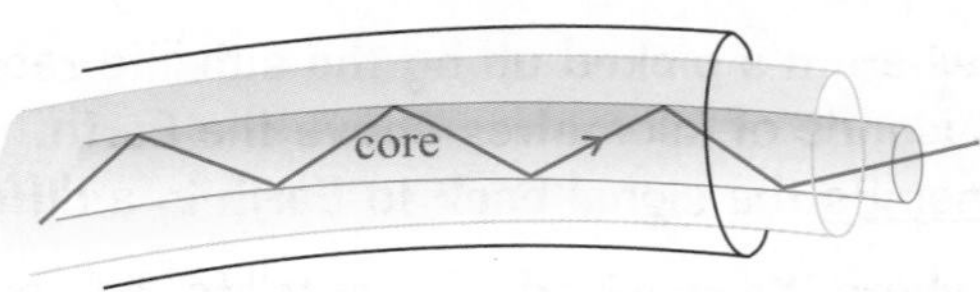

IR Can be Used to Monitor Temperature

1) Infrared radiation is also known as heat radiation. It's given out by hot objects — and the hotter the object, the more IR radiation it gives out.
2) This means infrared can be used to monitor temperatures. For example, heat loss through a house's uninsulated roof can be detected using infrared sensors, and security systems detect body heat.

3) Infrared is also detected by night-vision equipment. The equipment turns it into an electrical signal, which is displayed on a screen as a picture. The hotter an object is, the brighter it appears. Police and the military use this to spot baddies running away, like you've seen on TV.

IR Has Many Other Uses Around the Home

1) Infrared radiation can be used for cooking, e.g. in grills and toasters.
2) It can also be used to transmit information between mobile phones or computers — but only over short distances.

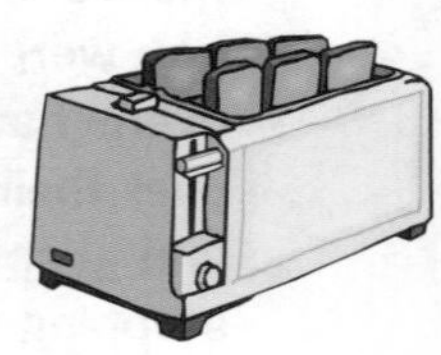

Visible Light is Useful for Photography

It sounds pretty obvious, but photography would be kinda tricky without visible light.

1) Cameras use a lens to focus visible light onto a light-sensitive film or electronic sensor.
2) The lens aperture controls how much light enters the camera (like the pupil in an eye).
3) The shutter speed determines how long the film or sensor is exposed to the light.
4) By varying the aperture and shutter speed (and also the sensitivity of the film or the sensor), a photographer can capture as much or as little light as they want in their photograph.

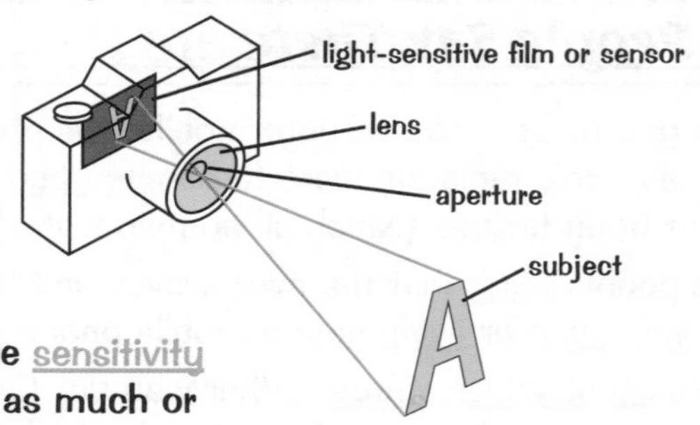

Don't lose control of your sensors — this page isn't remotely hard...

Because infrared technology is relatively cheap and cheerful we can afford to use it to make our lives a bit easier (and safer) around the home. Bad news for criminals. Remember — crime doesn't pay, revision does. Fact.

X-rays and Gamma Rays

Generally, high-energy EM radiation (like X-rays and gamma rays) is more harmful than low-energy radiation.

Some EM Radiation Causes Ionisation

1) All substances are made of atoms and molecules.
2) When radiation hits an atom or molecule, it sometimes has enough energy to remove an electron and change the atom or molecule. The changed atom is charged and is called an ion. This process is called ionisation:

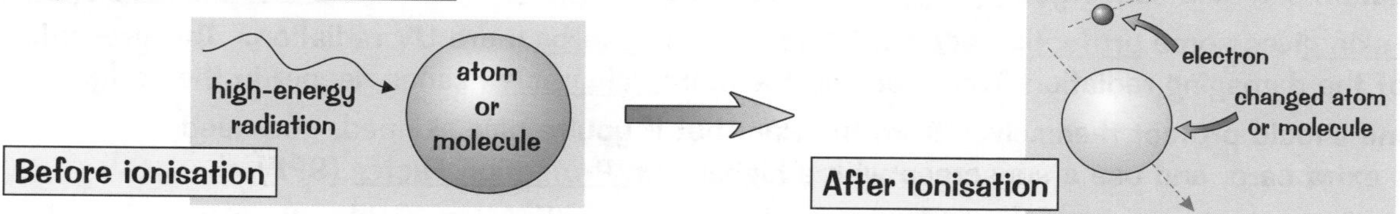

3) The changed atoms or molecules can go on to initiate (start) other chemical reactions.
4) It takes a lot of energy to remove an electron from an atom or molecule. So only the types of radiation with high enough energy can cause ionisation — ultraviolet, X-rays and gamma rays. These types of electromagnetic radiation are called ionising radiation.
5) Some substances (radioactive materials) emit ionising gamma radiation all the time (p.72).

Ionisation is Dangerous if it Happens in Your Cells

1) In the cells in your body, there are many important molecules, including DNA molecules.
2) If your cells are exposed to ionising radiation, the damage to DNA molecules can cause mutations, and the cells might start dividing over and over again, without stopping — this is cancer.
3) Very high doses of radiation can kill your cells altogether — this is what happens in 'radiation sickness'.
4) The longer you're exposed to the radiation the more damage it causes.

X-Rays are Used to Look Inside Objects

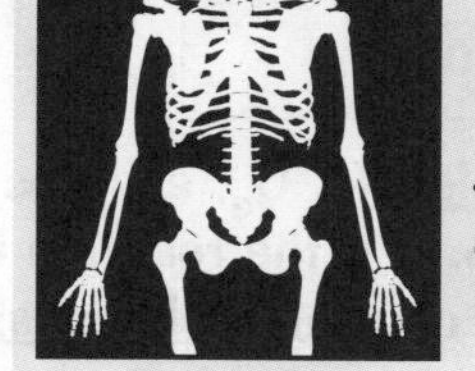

The brighter bits are where fewer X-rays get through. This is a negative image. The plate starts off all white.

1) Radiographers in hospitals take X-ray 'photographs' of people to see if they have any broken bones.
2) X-rays pass easily through flesh but not so easily through denser material like bones or metal. So it's the amount of radiation that's absorbed (or not absorbed) that gives you an X-ray image.
3) X-rays can cause cancer, so radiographers wear lead aprons and stand behind a lead screen or leave the room to keep their exposure to X-rays to a minimum.
4) Airport security use X-rays to scan luggage to check for suspicious-looking objects.
5) Some airports now use X-ray scanners on passengers to look for concealed weapons or explosives — low-level X-rays are used so they aren't as harmful as the X-rays used in hospitals.

Gamma Rays can be Used in Medicine

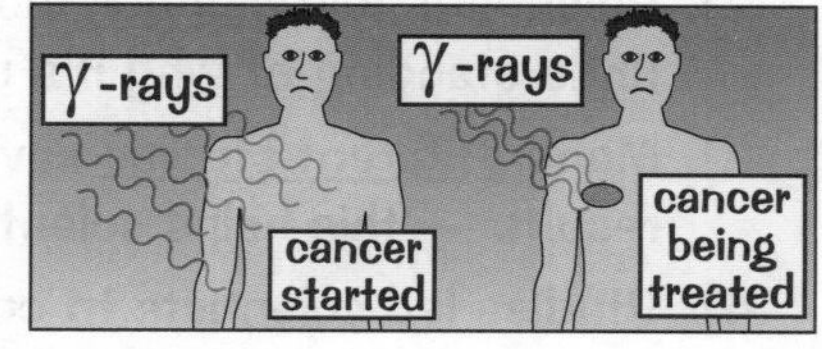

1) High doses of gamma (γ) rays can be used to treat cancers (see p.76)
2) Gamma rays can also be used to diagnose cancer. A radioactive isotope is injected into the patient — a gamma camera is then used to detect where the radioactive isotope travels in the body. This creates an image which can then be used to detect where there might be cancer (see p.76 for more).

Don't lie to an X-ray — they can see right through you...

X-rays and gamma rays can be harmful, but really useful too — as long as you're not exposed to them too much.

UV Radiation and Ozone

Ultraviolet (UV) radiation also causes ionisation. The Sun emits loads of UV, but the ozone layer can absorb it.

Ultraviolet Radiation Causes Skin Cancer

1) If you spend a lot of time in the sun, you can expect to get a tan and maybe sunburn.
2) But the more time you spend in the sun, the more chance you also have of getting skin cancer. This is because the Sun's rays include ultraviolet radiation which damages the DNA in your cells.
3) UV radiation can also cause you eye problems, such as cataracts, as well as premature skin aging (eek!).
4) Darker skin gives some protection against UV rays — it absorbs more UV radiation. This prevents some of the damaging radiation from reaching the more vulnerable tissues deeper in the body.
5) Everyone should protect themselves from the Sun, but if you're pale skinned, you need to take extra care, and use a sunscreen with a higher Sun Protection Factor (SPF).
6) An SPF of 15 means you can spend 15 times as long as you otherwise could in the sun without burning.
7) We're kept informed of the risks of exposure to UV — research into its effects is made public through the media and advertising, and the government tells people how to keep safe to improve public health.
8) It's not just exposure to the Sun that's a problem — we are now being warned of the risks of prolonged use of sunbeds too. Tanning salons have time limits to make sure people are not over-exposed.

The Ozone Layer Protects Us from UV Radiation

1) There's a layer of ozone high up in the Earth's atmosphere which absorbs some of the UV rays from the Sun — so it reduces the amount of UV radiation reaching the Earth's surface.
2) Ozone is a form of oxygen. An ozone molecule is just three oxygen atoms joined together — O_3. It's formed like this:

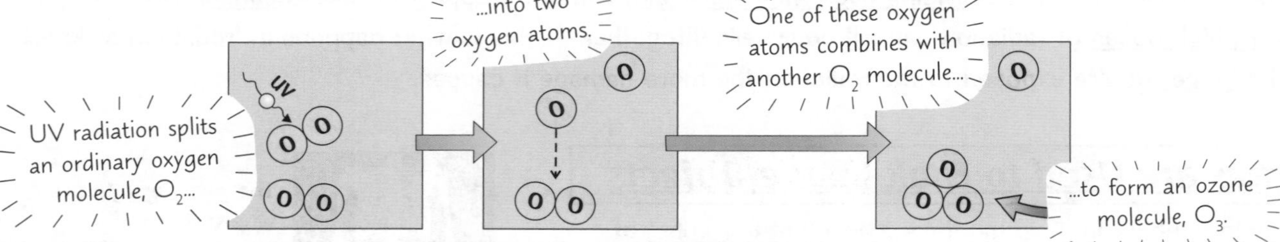

3) When an ozone molecule, O_3, absorbs more UV radiation, it splits into O_2 and O again. So the reaction is reversible (can go forwards and backwards) — causing a chemical change each time.

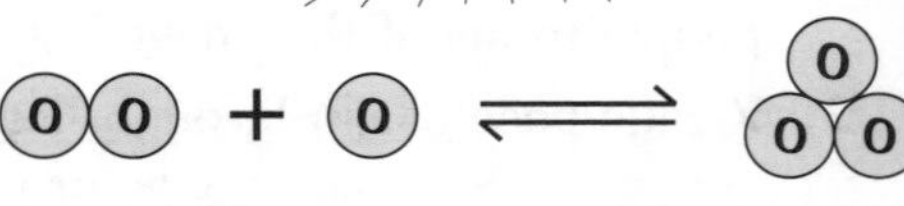

4) Recently, the ozone layer has got thinner because of pollution from CFCs — these are gases which react with ozone molecules and break them up. This depletion of the ozone layer allows more UV rays to reach us at the surface of the Earth (which, as you know, can be a danger to our health).

There's a Hole in the Ozone Layer over Antarctica

1) In winter, special weather effects cause the concentration of ozone over Antarctica to drop dramatically. It increases again in spring, but the winter concentration has been dropping. The low concentration looks like a 'hole' on satellite images.
2) Scientists now monitor the ozone concentration very closely to get a better understanding of why it's decreasing, and how to prevent further depletion.
3) Many different studies have been carried out internationally, using different equipment, to get accurate results — this helps scientists to be confident that their hypotheses and predictions are correct.
4) Studies led scientists to confirm that CFCs were causing the depletion of the ozone layer, so the international community banned them. We used to use CFCs all the time but now international bans and restrictions on CFC use have been put in place because of their environmental impact.

Use protection — wear a hat...

CFCs are man-made chemicals that were developed for use in aerosol sprays and refrigerators. So now you know.

The Greenhouse Effect

The atmosphere keeps us warm by trapping heat.

Some Radiation from the Sun Passes Through the Atmosphere

1) The Earth is surrounded by an atmosphere made up of various gases — the air.
2) The gases in the atmosphere filter out certain types of radiation from the Sun — they absorb or reflect radiation of certain wavelengths (infrared).
3) However, some wavelengths of radiation — mainly visible light and some radio waves — pass through the atmosphere quite easily.

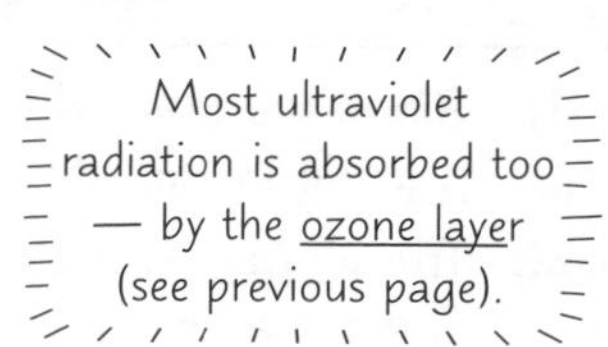

The Greenhouse Effect Helps Regulate Earth's Temperature

1) The Earth absorbs short wavelength EM radiation from the Sun. This warms the Earth's surface up. The Earth then emits some of this EM radiation back out into space — this tends to cool us down.
2) Most of the radiation emitted from Earth is longer wavelength infrared radiation — heat.
3) A lot of this infrared radiation is absorbed by atmospheric gases, including carbon dioxide, methane and water vapour.
4) These gases then re-radiate heat in all directions, including back towards the Earth.
5) So the atmosphere acts as an insulating layer, stopping the Earth losing all its heat at night.
6) This is known as the 'greenhouse effect'. (In a greenhouse, the sun shines in and the glass helps keep some of the heat in.) Without the greenhouse gases (CO_2, methane, water vapour) in our atmosphere, the Earth would be a lot colder.

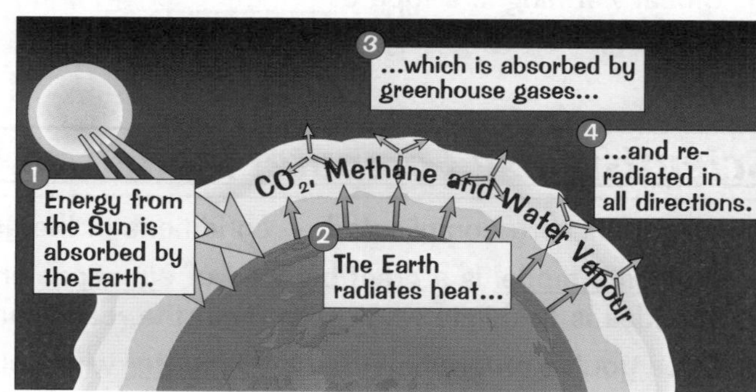

Humans are Causing an Increase in the Amount of Greenhouse Gases

Over the last 200 years or so, the concentration of greenhouse gases in the atmosphere has been increasing. This is because some of the sources of them are increasing, so more gases are being released:

Carbon Dioxide

People use more energy (e.g. travel more in cars) — which we get mainly from burning fossil fuels, which releases more carbon dioxide.

More land is needed for houses and food and the space is often made by chopping down and burning trees — fewer trees mean less CO_2 is absorbed, and burning releases more CO_2.

CO_2 also comes from natural sources — e.g. respiration in animals and plants, and volcanic eruptions can release it.

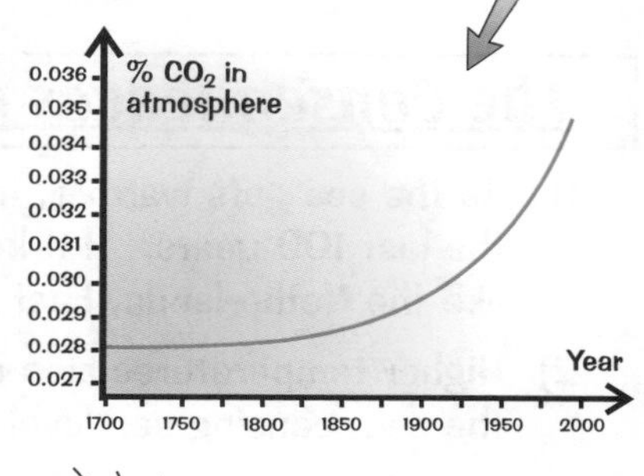

You might have to interpret data on greenhouse gases — there's more on interpreting data on p.7.

Methane

Cattle farming has increased to feed the growing population — cattle digestion produces methane, so the amount of methane is increasing.

Decaying waste in landfill sites produces methane — the amount of waste is increasing, causing, you guessed it, an increase in methane.

Methane is released naturally by volcanoes, wetlands and wild animals.

Water Vapour

Most water vapour comes from natural sources — mainly oceans, seas, rivers and lakes.
As global temperature increases (see next page), so could the amount of water vapour.
Power stations also produce water vapour, which can affect the amount in the local area.

A biologist, a chemist and a physicist walk into a greenhouse...

...it works out badly. Without the greenhouse effect the Earth would be pretty nippy. Brrrrr.

Global Warming and Climate Change

Without any 'greenhouse gases' in the atmosphere, the Earth would be about 30 °C colder than it is now. So we need the greenhouse effect — just not too much of it...

Upsetting the Greenhouse Effect Has Led to Global Warming

1) Since we started burning fossil fuels in a big way, the level of carbon dioxide in the atmosphere has increased (see previous page).
2) The global temperature has also risen during this time (global warming). There's a link between concentration of CO_2 and global temperature.
3) A lot of evidence shows that the rise in CO_2 level is causing global warming by increasing the greenhouse effect (see previous page).
4) So there's now a scientific consensus (general agreement) that humans are causing global warming.
5) Global warming is a type of climate change, and it also causes other types, e.g. changing weather patterns.

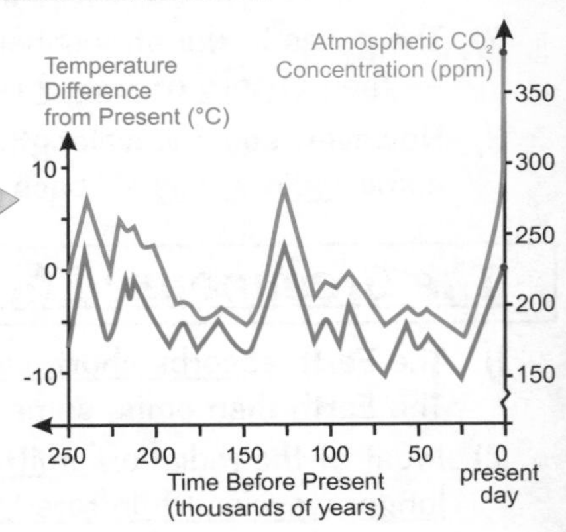

Scientists Use Computer Models to Understand Climate Change

The climate is very complicated — conditions in the atmosphere, oceans and land all affect one another.

1) A climate model is a great big load of equations linking these various parts of the climate system. The idea is to mimic what goes on in the real climate by doing calculations.
2) Once you've programmed a big computer with your equations, you need some data to start the calculations off. E.g. you might put in some data about temperature at the surface of the ocean in various places. The computer uses this data to work out, say, the speed and direction of ocean currents... then uses those results to work out air temperatures around the world.
3) Climate models can also be used to explain why the climate is changing now. We know that the Earth's climate varies naturally — changes in our orbit around the Sun cause ice ages, for instance. Climate modelling over the last few years has shown that natural changes don't explain the current 'global warming' — and that the increase in greenhouse gases due to human activity is the cause.

The Consequences of Global Warming Could be Pretty Serious

1) As the sea gets warmer, it expands, causing sea level to rise. Sea level has risen a little bit over the last 100 years. If it keeps rising it'll be bad news for people living in low-lying places like the Netherlands, East Anglia and the Maldives — they'd be flooded.
2) Higher temperatures also make ice melt. Water that's currently 'trapped' on land as ice runs into the sea, causing sea level to rise even more.
3) Global warming has changed weather patterns in many parts of the world. It's thought that many regions will suffer more extreme weather because of this, e.g. longer, hotter droughts. Hurricanes form over water that's warmer than 27 °C — so with more warm water, you'd expect more hurricanes.
4) The extra heat in the atmosphere will also increase convection (stronger winds) and result in more water vapour (more rain), causing more storms and floods.
5) Changing weather patterns also affect food production — some regions are now too dry to grow food, some too wet. This will get worse as temperature increases and weather patterns change more.

Be a climate model — go on a diet and solve lots of equations...

'Global warming' could mean that some parts of the world cool down. For instance, as ice melts, lots of cold fresh water will enter the sea and this could disrupt the ocean currents. This could be bad news for us in Britain — if the nice warm currents we get at the moment weaken, we'll be a lot colder.

Revision Summary for Section Two

More jolly questions which I know you're going to really enjoy. If you struggle with any of them, have another read through the section and give the questions another go.

1) Draw a diagram to illustrate the frequency, wavelength and amplitude of a wave.
2) Give the equation that links wave speed, frequency and wavelength.
3)* Find the speed of a wave with frequency 50 000 Hz and wavelength 0.3 cm.
4) a) Sketch a diagram of a ray of light being reflected in a plane mirror.
   b) Label the normal and the angles of incidence and reflection.
5) Draw a diagram showing a wave diffracting through a gap.
6) What size should the gap be in order to maximise diffraction?
   a) much larger than the wavelength
   b) the same size as the wavelength
   c) a bit bigger than the wavelength.
7) Why does light bend if it hits a boundary between air and water at an angle?
8) Why can't sound waves travel in space?
9) Are high frequency sound waves high-pitched or low-pitched?
10) Draw diagrams of an analogue and a digital signal and briefly explain the differences between them.
11) Why are digital signals better quality than analogue signals?
12) Sketch the EM spectrum with all its details. Put the lowest frequency waves on the left.
13) Explain why radio waves can be transmitted across long distances.
14) Describe how satellites are used for communication.
15) Briefly explain how microwaves cook food.
16) What type of wave do remote controls usually use?
17) Which two types of EM wave are commonly used to send signals along optical fibres?
18) Why is ionisation dangerous if it occurs in your cells?
19) Explain how both X-rays and gamma rays can be useful in hospitals.
20) Give three health problems that can be caused by exposure to UV radiation.
21) How is ozone made in the atmosphere?
22) Briefly describe what is meant by the 'greenhouse effect'.
23) What two effects does chopping down and burning trees have on the atmospheric carbon dioxide level?
24) What is global warming?
25) What's causing global warming? How do we know this?
26) Give two possible consequences of global warming.

* Answers on page 140.

Velocity and Acceleration

Speed, velocity, acceleration... no doubt you've heard these words being bandied about during dinner parties. If you've ever felt out of your depth when talk turns to the differences between them, this page is for you...

Speed and Velocity are Both: HOW FAST YOU'RE GOING

Speed and velocity are both measured in m/s (or km/h or mph). They both simply say how fast you're going, but there's a subtle difference between them which you need to know:

> SPEED is just how fast you're going (e.g. 30 mph or 20 m/s) with no regard to the direction.
>
> VELOCITY however must also have the DIRECTION specified, e.g. 30 mph north or 20 m/s, 060°. The distance in a particular direction is called the DISPLACEMENT.

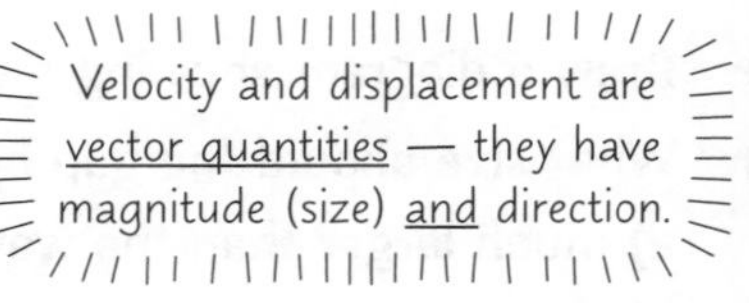

You can have negative velocities. If a car travelling at 20 m/s then turns around to go in the opposite direction, the speed is still 20 m/s but the velocity becomes –20 m/s.

Speed, Distance and Time — the Formula:

You really ought to get pretty slick with this very easy equation, it pops up a lot...

$$\text{Speed} = \frac{\text{Distance}}{\text{Time}}$$

EXAMPLE: A cat skulks 20 m in 35 s. Find: a) its speed, b) how long it takes to skulk 75 m.

ANSWER: Using the formula triangle: a) $s = d/t = 20/35 = \underline{0.57 \text{ m/s}}$

b) $t = d/s = 75/0.57 = 132 \text{ s} = \underline{2 \text{ min } 12 \text{ s}}$

A lot of the time we tend to use the words "speed" and "velocity" interchangeably. But if you're asked to calculate a velocity in the exam, don't forget to state a direction.

Acceleration is How Quickly Velocity is Changing

Acceleration is definitely not the same as velocity or speed.

1) Acceleration is how quickly the velocity is changing.
2) This change in velocity can be a CHANGE IN SPEED or a CHANGE IN DIRECTION or both. (You only have to worry about the change in speed bit for calculations.)

BUT, acceleration is a vector quantity like velocity — it has magnitude and direction.

Acceleration — The Formula:

$$\text{Acceleration} = \frac{\text{Change in Velocity}}{\text{Time taken}}$$

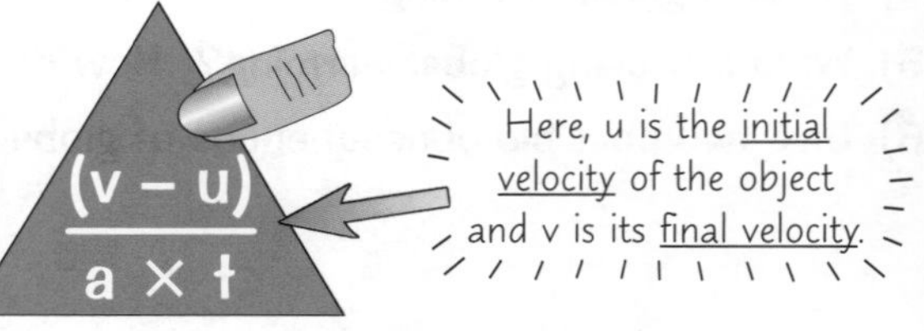

There are two tricky things with this equation. First there's the '(v – u)', which means working out the 'change in velocity', as shown in the example below, rather than just putting a simple value for velocity or speed in. Secondly there's the unit of acceleration, which is m/s². (Don't get confused with the units for velocity, m/s).

(v – u) could also be written Δv — it just means "change in v".

EXAMPLE: A skulking cat accelerates from 2 m/s to 6 m/s in 5.6 s. Find its acceleration.

ANSWER: Using the formula triangle: $a = (v - u) / t = (6 - 2) / 5.6 = 4 \div 5.6 = \underline{0.71 \text{ m/s}^2}$

They say a change in velocity is as good as a rest...

Lots of facts to remember there, but it's all important stuff. Make sure you've learnt it all before you move on. Remember — displacement, velocity and acceleration are all vector quantities because they have size and direction.

D-T and V-T Graphs

Make sure you learn all these details real good. Make sure you can distinguish between the two graphs too.

Distance-Time Graphs

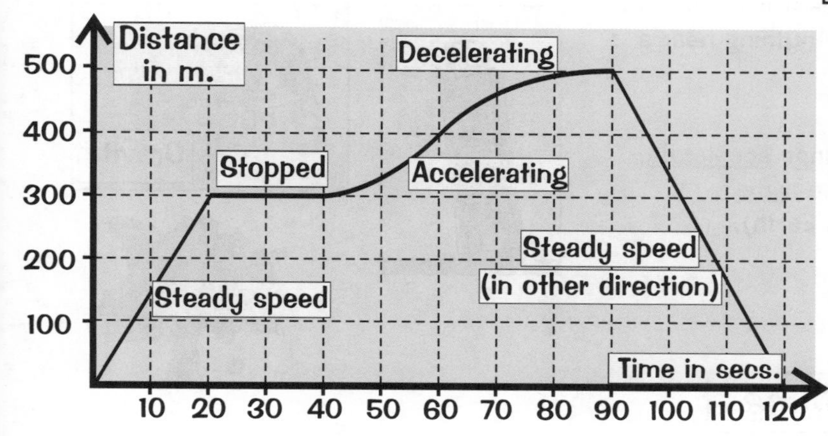

Very Important Notes:

1) Gradient = speed.
2) Flat sections are where it's stopped.
3) The steeper the graph, the faster it's going.
4) Downhill sections mean it's going back toward its starting point.
5) Curves represent acceleration or deceleration.
6) A steepening curve means it's speeding up (increasing gradient).
7) A levelling off curve means it's slowing down (decreasing gradient).

Calculating Speed from a Distance-Time Graph — It's Just the Gradient

For example, the speed of the return section of the graph is:

$$\text{Speed} = \text{gradient} = \frac{\text{vertical}}{\text{horizontal}} = \frac{500}{30} = 16.7 \text{ m/s}$$

This is just the speed equation (p.40).

Don't forget that you have to use the scales of the axes to work out the gradient. Don't measure in cm!

Velocity-Time Graphs

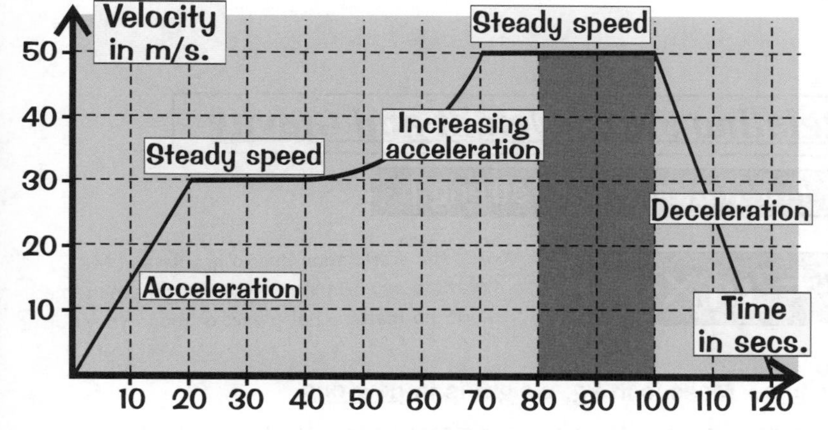

Very Important Notes:

1) Gradient = acceleration.
2) Flat sections represent steady speed.
3) The steeper the graph, the greater the acceleration or deceleration.
4) Uphill sections (/) are acceleration.
5) Downhill sections (\) are deceleration.
6) The area under any section of the graph (or all of it) is equal to the distance travelled in that time interval.
7) A curve means changing acceleration.

Calculating Acceleration and Distance from a Velocity-Time Graph

1) The acceleration represented by the first section of the graph is:

$$\text{Acceleration} = \text{gradient} = \frac{\text{vertical}}{\text{horizontal}} = \frac{30}{20} = 1.5 \text{ m/s}^2$$

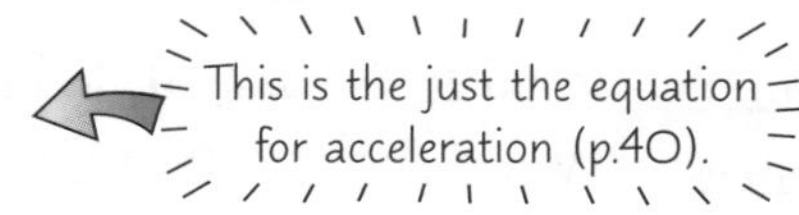

2) The distance travelled in any time interval is equal to the area under the graph. For example, the distance travelled between t = 80 s and t = 100 s is equal to the shaded area which is equal to 1000 m. (But we can only use the method for uniform (constant or steady) acceleration).

Understanding motion graphs — it can be a real uphill struggle...

The tricky thing about these two types of graph is they can look pretty much the same but represent totally different kinds of motion. When you read from a motion graph, check the axis labels carefully so you know which type it is.

Weight, Mass and Gravity

Now for something a bit more attractive — the force of gravity. Enjoy...

Gravitational Force is the Force of Attraction Between All Masses

Gravity attracts all masses, but you only notice it when one of the masses is really really big, e.g. a planet. Anything near a planet or star is attracted to it very strongly.

This has two important effects:

1) On the surface of a planet, it makes all things accelerate (see p.40) towards the ground (all with the same acceleration, g, which is about 10 m/s^2 on Earth).
2) It gives everything a weight.

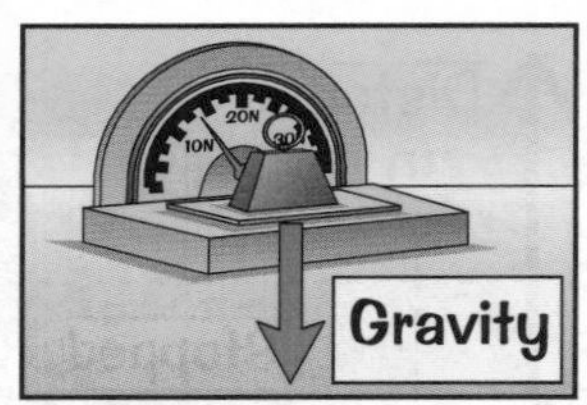

Weight and Mass are Not the Same

1) Mass is just the amount of 'stuff' in an object. For any given object this will have the same value anywhere in the Universe.
2) Weight is caused by the pull of the gravitational force. In most questions the weight of an object is just the force of gravity pulling it towards the centre of the Earth.
3) An object has the same mass whether it's on Earth or on the Moon — but its weight will be different. A 1 kg mass will weigh less on the Moon (about 1.6 N) than it does on Earth (about 10 N), simply because the gravitational force pulling on it is less.
4) Weight is a force measured in newtons. It's measured using a spring balance or newton meter. Mass is not a force. It's measured in kilograms with a mass balance (an old-fashioned pair of balancing scales).

The Very Important Formula Relating Mass, Weight and Gravity

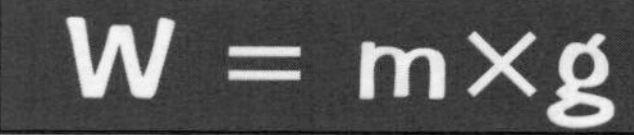

$$W = m \times g$$

The acceleration due to gravity and the gravitational field strength are always the same value, no matter what planet or moon you're on.

1) Remember, weight and mass are not the same. Mass is in kg, weight is in newtons.
2) The letter "g" represents the strength of the gravity and its value is different for different planets. On Earth g ≈ 10 N/kg. On the Moon, where the gravity is weaker, g is only about 1.6 N/kg.
3) This formula is hideously easy to use:

Example: What is the weight, in newtons, of a 5 kg mass, both on Earth and on the Moon?

Answer: "W = m × g". On Earth: W = 5 × 10 = 50 N (The weight of the 5 kg mass is 50 N.)
On the Moon: W = 5 × 1.6 = 8 N (The weight of the 5 kg mass is 8 N.)

See what I mean. Hideously easy — as long as you've learnt what all the letters mean.

I don't think you understand the gravity of this situation...

The difference between weight and mass can be tricky to get your head around, but it's well important. Weight is the force of gravity acting on a mass, and mass is the amount of stuff, measured in kg. Sorted? Good.

Resultant Forces

Gravity isn't the only force in town — there are other forces such as driving forces or air resistance. What you need to be able to work out is how all these forces add up together.

Resultant Force is the Overall Force on a Point or Object

The notion of resultant force is a really important one for you to get your head round:

1) In most real situations there are at least two forces acting on an object along any direction.
2) The overall effect of these forces will decide the motion of the object — whether it will accelerate, decelerate or stay at a steady speed.
3) If you have a number of forces acting at a single point, you can replace them with a single force (so long as the single force has the same effect on the motion as the original forces acting all together).
4) If the forces all act along the same line (they're all parallel and act in the same or the opposite direction), the overall effect is found by just adding or subtracting them.
5) The overall force you get is called the resultant force.

Example: Stationary Teapot — All Forces Balance

1) The force of GRAVITY (or weight) is acting downwards.
2) This causes a REACTION FORCE (see p.45) from the surface pushing up on the object.
3) This is the only way it can be in BALANCE.
4) Without a reaction force, it would accelerate downwards due to the pull of gravity.
5) The resultant force on the teapot is zero: 10 N – 10 N = 0 N.

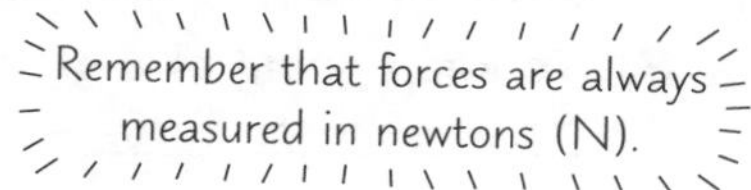

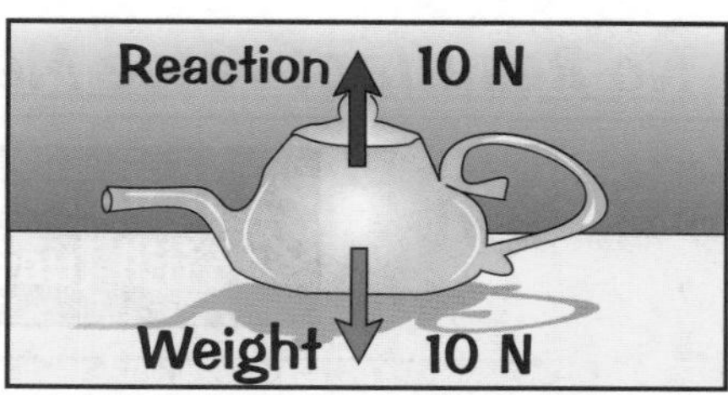

1) The length of the arrow shows the size of the force.
2) The direction of the arrow shows the direction of the force (didn't see that one coming, did you...).
3) If the arrows come in opposite pairs, and they're all the same size, then the forces are balanced.

A Resultant Force Means a Change in Velocity

1) If there is a resultant force acting on an object, then the object will change its state of rest or motion.
2) In other words it causes a change in the object's velocity.

You Should be Able to Find the Resultant Force Acting in a Straight Line

EXAMPLE: Benny is cruising along to Las Vegas in his vintage sports car. He applies a driving force of 1000 N, but has to overcome air resistance of 600 N. What is the resultant force? Will the car's velocity change?

ANSWER: Say that the forces pointing to the left are pointing in the positive direction.
The resultant force = 1000 N – 600 N = 400 N to the left.
If there is a resultant force then there is always an acceleration, so Benny's velocity will change. Viva Las Vegas.

And you're moving forward — what a result...

Resultant forces are just about adding and subtracting really — the trick is to make sure you've accounted for everything. Next up, some of the thrilling physics you can understand once you have resultant forces figured out.

Forces and Acceleration

Around about the time of the Great Plague in the 1660s, a chap called Isaac Newton worked out his Laws of Motion. At first they might seem kind of obscure or irrelevant, but to be perfectly blunt, if you can't understand this page then you'll never understand forces and motion.

An Object Needs a Force to Start Moving

If the resultant force on a stationary object is zero, the object will remain stationary.

Things don't just start moving on their own, there has to be a resultant force (see p.43) to get them started.

No Resultant Force Means No Change in Velocity

If there is no resultant force on a moving object it'll just carry on moving at the same velocity.

1) When a train or car or bus or anything else is moving at a constant velocity then the forces on it must all be balanced.
2) Never let yourself entertain the ridiculous idea that things need a constant overall force to keep them moving — NO NO NO NO NO NO!
3) To keep going at a steady speed, there must be zero resultant force — and don't you forget it.

A Resultant Force Means Acceleration

If there is a non-zero resultant force, then the object will accelerate in the direction of the force.

1) A non-zero resultant force will always produce acceleration (or deceleration).
2) This "acceleration" can take five different forms: Starting, stopping, speeding up, slowing down and changing direction.
3) On a force diagram, the arrows will be unequal:

Don't ever say: "If something's moving there must be an overall resultant force acting on it".
Not so. If there's an overall force it will always accelerate.
You get steady speed when there is zero resultant force.
I wonder how many times I need to say that same thing before you remember it?

Steady Speed Bus Tours Ltd. — providing consistent service since 1926...

Objects in space don't need a driving force to keep travelling at a steady speed — it's only because of air resistance and friction that we do. A steady speed means that there is zero resultant force.

Forces and Acceleration

More fun stuff on forces and acceleration. The big equation here is F = ma.
Remember that the F is always the resultant force — that's important too.

A Non-Zero Resultant Force Produces an Acceleration

Any resultant force will produce acceleration, and this is the formula for it:

$$F = ma \quad \text{or} \quad a = F/m$$

m = mass in kilograms (kg)
a = acceleration in metres per second squared (m/s^2)
F is the resultant force in newtons (N)

EXAMPLE: A car of mass of 1750 kg has an engine which provides a driving force of 5200 N. At 70 mph the drag force acting on the car is 5150 N. Find its acceleration a) when first setting off from rest b) at 70 mph.

ANSWER: 1) First draw a force diagram for both cases (no need to show the vertical forces):

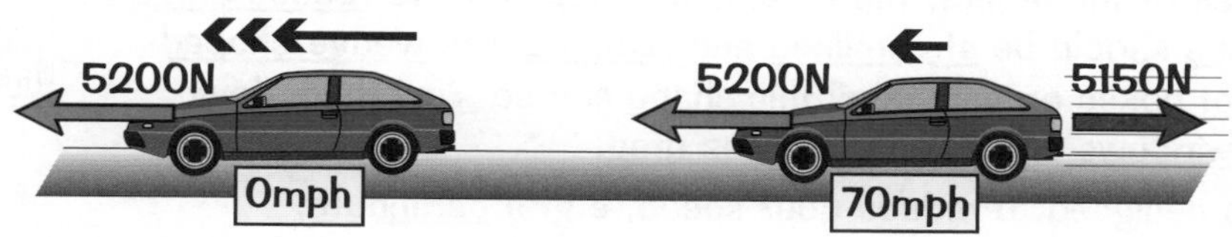

2) Work out the resultant force and acceleration of the car in each case.

Resultant force = 5200 N
a = F/m = 5200 ÷ 1750 = 3.0 m/s^2

Resultant force = 5200 – 5150 = 50 N
a = F/m = 50 ÷ 1750 = 0.03 m/s^2

Reaction Forces are Equal and Opposite

When two objects interact, the forces they exert on each other are equal and opposite.

These two forces are called an 'interaction pair'.

1) That means if you push something, say a shopping trolley, the trolley will push back against you, just as hard.
2) And as soon as you stop pushing, so does the trolley. Kinda clever really.
3) So far so good. The slightly tricky thing to get your head round is this — if the forces are always equal, how does anything ever go anywhere? The important thing to remember is that the two forces are acting on different objects. Think about a pair of ice skaters:

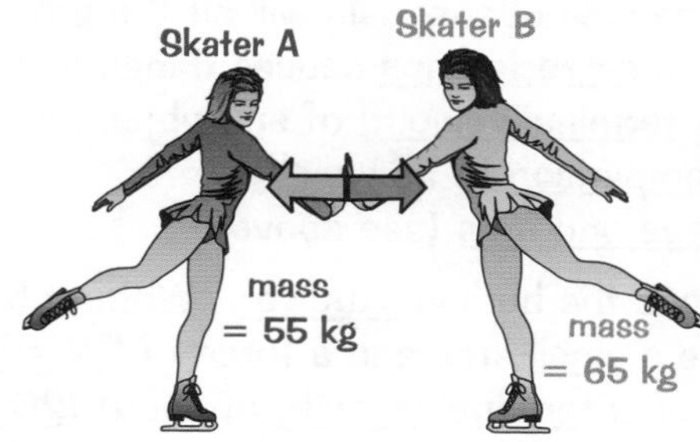

When skater A pushes on skater B (the 'action' force), she feels an equal and opposite force from skater B's hand (the 'reaction' force). Both skaters feel the same sized force, in opposite directions, and so accelerate away from each other.

Skater A will be accelerated more than skater B, though, because she has a smaller mass — remember a = F/m.

4) It's the same sort of thing when you go swimming. You push back against the water with your arms and legs, and the water pushes you forwards with an equal-sized force in the opposite direction.

I have a reaction to forces — they bring me out in a rash...

This is the real deal. Like... proper Physics. It was pretty fantastic at the time it was discovered — suddenly people understood how forces and motion worked, they could work out the orbits of planets and everything.

Frictional Forces and Terminal Velocity

Imagine a world without friction — you'd be sliding around all over the place. Weeeeeeeee.... Ouch.

Friction Will Slow Things Down

1) When an object is moving (or trying to move) friction acts in the direction that opposes movement.
2) The frictional force will match the size of the force trying to move it, up to a point — after this the friction will be less than the other force and the object will move.
3) Friction will act to make the moving object slow down and stop.
4) So to travel at a steady speed, things always need a driving force to overcome the friction.
5) Friction occurs in three main ways:

   a) FRICTION BETWEEN SOLID SURFACES WHICH ARE GRIPPING (static friction)

   b) FRICTION BETWEEN SOLID SURFACES WHICH ARE SLIDING PAST EACH OTHER

sliding friction

   c) RESISTANCE OR "DRAG" FROM FLUIDS (LIQUIDS OR GASES, e.g. AIR)

   The larger the area of the object, the greater the drag. So, to reduce drag, the area and shape should be streamlined and reduced, like wedge-shaped sports cars. Roof boxes on cars spoil this shape and so slow them down. Driving with the windows open also increases drag.

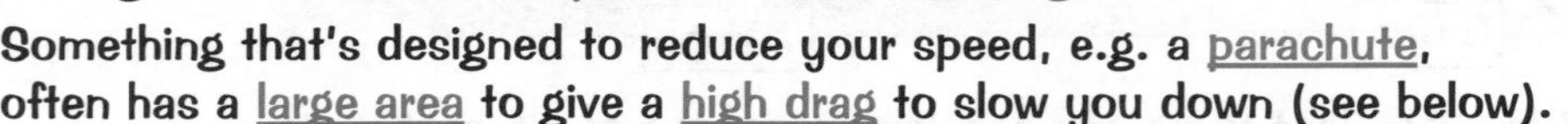

   Something that's designed to reduce your speed, e.g. a parachute, often has a large area to give a high drag to slow you down (see below).

   In a fluid: FRICTION (DRAG) ALWAYS INCREASES AS THE SPEED INCREASES — and don't forget it.

Moving Objects Can Reach a Terminal Velocity

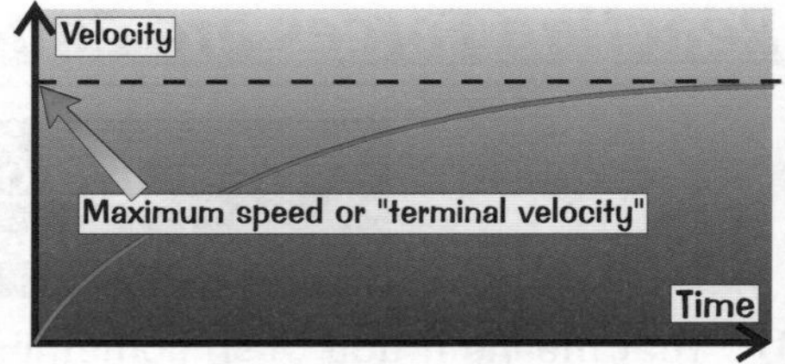

1) When objects first set off they have much more force accelerating them than resistance slowing them down.
2) As the speed increases, the resistance increases as well.
3) This gradually reduces the acceleration until the resistance force (friction or drag) is equal to the accelerating force (weight or thrust) so it can't accelerate any more. The forces are balanced.
4) It will have reached its maximum speed or terminal velocity.

The Terminal Velocity of Falling Objects Depends on their Shape and Area

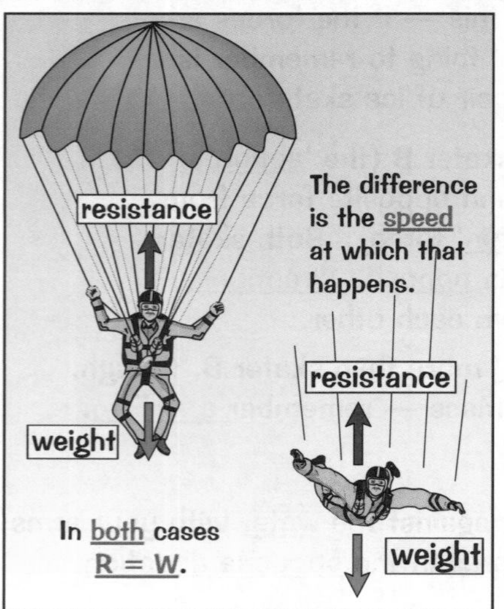

The accelerating force acting on all falling objects is gravity and it would make them all fall at the same rate, if it wasn't for air resistance. This means that on the Moon, where there's no air, hamsters and feathers dropped simultaneously will hit the ground together. However, on Earth, air resistance causes things to fall at different speeds, and the terminal velocity of any object is determined by its drag in comparison to its weight. The drag depends on its shape and area (see above).

The most important example is the human skydiver. Without his parachute open he has quite a small area and a force of "W = mg" pulling him down. He reaches a terminal velocity of about 120 mph. But with the parachute open, there's much more air resistance (at any given speed) and still only the same force "W = mg" pulling him down. This means his terminal velocity comes right down to about 15 mph, which is a safe speed to hit the ground at.

Learning about air resistance — it can be a real drag...

There are a few really important things on this page. 1) When you fall through a fluid, there's a frictional force (drag), 2) frictional force increases with speed, so 3) you eventually reach terminal velocity.

Stopping Distances

And now a page on stopping distances. This may seem a bit out of kilter with the rest of the section, but it's a real world application of the physics of forces. See, I told you it was useful... and fun... right?

Many Factors Affect Your Total Stopping Distance

1) Looking at things simply — if you need to stop in a given distance, then the faster a vehicle's going, the bigger braking force it'll need.
2) Likewise, for any given braking force, the faster you're going, the greater your stopping distance. But in real life it's not quite that simple — if your maximum braking force isn't enough, you'll go further before you stop.
3) The total stopping distance of a vehicle is the distance covered in the time between the driver first spotting a hazard and the vehicle coming to a complete stop.
4) The stopping distance is the sum of the thinking distance and the braking distance.

1) Thinking Distance

"The distance the vehicle travels during the driver's reaction time".

The reaction time is the time between the driver spotting a hazard and taking action.

It's affected by two main factors:

a) How fast you're going — Obviously. Whatever your reaction time, the faster you're going, the further you'll go.

b) How dopey you are — This is affected by tiredness, drugs, alcohol and a careless blasé attitude.

Bad visibility and distractions can also be a major factor in accidents — lashing rain, messing about with the radio, bright oncoming lights, etc. might mean that a driver doesn't notice a hazard until they're quite close to it. It doesn't affect your thinking distance, but you start thinking about stopping nearer to the hazard, and so you're more likely to crash.

The figures below for typical stopping distances are from the Highway Code. It's frightening to see just how far it takes to stop when you're going at 70 mph.

	30 mph	50 mph	70 mph
Thinking distance	9 m	15 m	21 m
Braking distance	14 m	38 m	75 m
Total	6 car lengths	13 car lengths	24 car lengths

2) Braking Distance

"The distance the car travels under the braking force".

It's affected by five main factors:

a) How fast you're going — The faster you're going, the further it takes to stop.

b) How good your brakes are — All brakes must be checked and maintained regularly. Worn or faulty brakes will let you down catastrophically just when you need them the most, i.e. in an emergency.

c) The mass of your vehicle — With the same brakes, a heavily laden vehicle takes longer to stop.

d) How good the tyres are — Tyres should have a minimum tread depth of 1.6 mm in order to be able to get rid of the water in wet conditions. Leaves, diesel spills and muck on the road can greatly increase the braking distance, and cause the car to skid too.

e) How good the grip is — This depends on three things: 1) road surface, 2) weather conditions, 3) tyres.

Wet or icy roads are always much more slippy than dry roads, but often you only discover this when you try to brake hard. You don't have as much grip, so you travel further before stopping.

Stop right there — and learn this page...

Without tread, a tyre will simply ride on a layer of water and skid very easily. This is called "aquaplaning" and isn't nearly as cool as it sounds. Snow and ice are also very hazardous because it is difficult for the tyres to get a grip.

Momentum

A large lorry being driven very fast is going to be a lot harder to stop than a granny on a bicycle out for a Sunday afternoon ride — that's momentum for you.

Momentum = Mass × Velocity

momentum / mass × velocity

1) The greater the mass of an object and the greater its velocity, the more momentum the object has.
2) Momentum is a vector quantity — it has size and direction (like velocity, but not speed, see p.40).

Momentum (kg m/s) = Mass (kg) × Velocity (m/s)

Momentum Before = Momentum After

Momentum is conserved when no external forces act, i.e. the total momentum after is the same as it was before. This is particularly obvious when you have a linear system — when the forces are working along the same line.

Example:

Two skaters approach each other, collide and move off together as shown. At what velocity do they move after the collision?

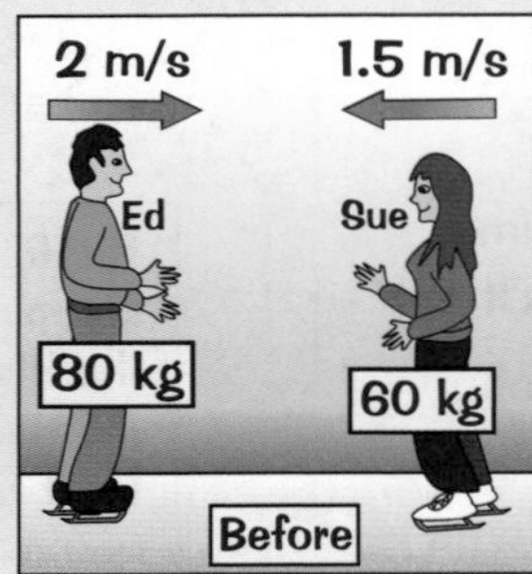

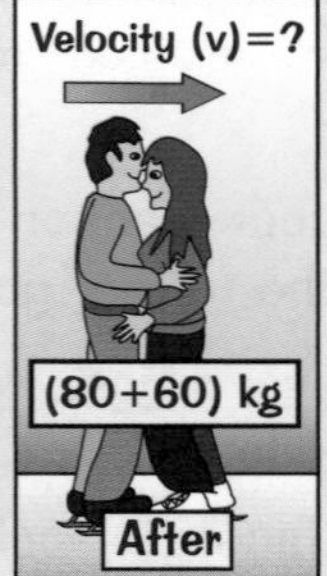

1) Choose which direction is positive. I'll say "positive" means "to the right".
2) Total momentum before collision
   = momentum of Ed + momentum of Sue
   = {80 × 2} + {60 × (–1.5)} = 70 kg m/s
3) Total momentum after collision
   = momentum of Ed and Sue together
   = 140 × v
4) So 140v = 70, i.e. v = 0.5 m/s to the right.

Forces Cause Changes in Momentum

1) When a force acts on an object, it causes a change in momentum.

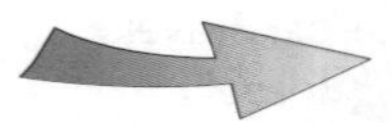

$$\text{Force (N)} = \frac{\text{Change in momentum (kg m/s)}}{\text{Time (s)}}$$

2) A larger force means a faster change of momentum (and so a greater acceleration, see p.45).
3) Likewise, if someone's momentum changes very quickly (like in a car crash), the forces on the body will be very large, and more likely to cause injury.
4) This is why cars are designed with protective features to slow people down over a longer time when they have a crash — the longer it takes for a change in momentum, the smaller the force (see page 52).

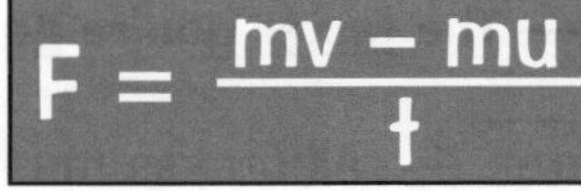

$$F = \frac{mv - mu}{t}$$

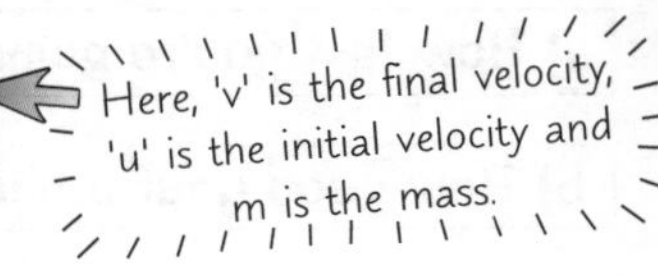

(mv – mu) can also be written ΔM — it just means 'change in momentum'.

EXAMPLE: A rock with mass 1 kg is travelling through space at 15 m/s. A comet hits the rock, exerting a force of 2500 N on it for 0.7 seconds. Calculate a) the rock's initial momentum, and b) the change in its momentum resulting from the impact.

ANSWER: a) Momentum = mass × velocity = 1 × 15 = 15 kg m/s
b) Rearranging the formula,
Change of momentum = force × time = 2500 × 0.7 = 1750 kg m/s.

Learn this stuff — it'll only take a moment... um...

Momentum's a pretty fundamental bit of Physics — so make sure you learn all the stuff on this page properly. Momentum depends on mass and velocity, and a force can result in a change in momentum. Safety features on cars work by slowing down the change in momentum — there's more on this on page 52.

Work and Power

In Physics, "work done" means something special — it's got its own formula and everything.

When a force moves an object, energy is transferred and work is done.

That statement sounds far more complicated than it needs to. Try this:

1) Whenever something moves, something else is providing some sort of 'effort' to move it.
2) The thing putting the effort in needs a supply of energy (like fuel or food or electricity etc.).
3) It then does 'work' by moving the object — and one way or another it transfers the energy it receives (as fuel etc.) into other forms (see p.16).

4) Whether this energy is transferred 'usefully' (e.g. by lifting a load) or is 'wasted' (e.g. lost as heat through friction), you can still say that 'work is done'. Just like Batman and Bruce Wayne, 'work done' and 'energy transferred' are indeed 'one and the same'. (And they're both given in joules.)

It's Just Another Trivial Formula:

Work Done = Force × Distance moved in the direction of the force

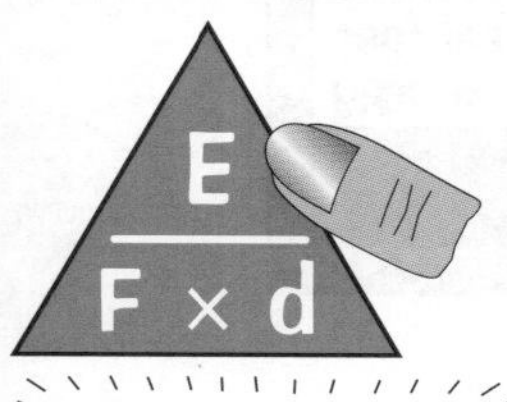

You might also see W used to represent work done in this equation.

Whether the force is friction or weight or tension in a rope, it's always the same. To find how much energy has been transferred (in joules), you just multiply the force in N by the distance moved in the direction of the force, measured in metres. Easy as that. I'll show you...

EXAMPLE: Some hooligan kids drag an old tractor tyre 5 m over rough ground. They pull with a total force of 340 N. Find the energy transferred.

ANSWER: Energy transferred is work done, so:
$E = F \times d = 340 \times 5 = 1700$ J. Phew — easy peasy isn't it?

Power is the "Rate of Doing Work" — i.e. How Much per Second

Power is not the same thing as force, nor energy. A powerful machine is not necessarily one which can exert a strong force (though it usually ends up that way). A powerful machine is one which transfers a lot of energy in a short space of time. This is the very easy formula for power:

$$\text{Power} = \frac{\text{Work done}}{\text{Time taken}}$$

EXAMPLE: A motor transfers 4.8 kJ of useful energy in 2 minutes. Find its power output.

ANSWER: $P = E / t = 4800/120 = 40$ W (or 40 J/s)
(Note that the kJ had to be turned into J, and the minutes into seconds.)

1 kJ = 1000 J

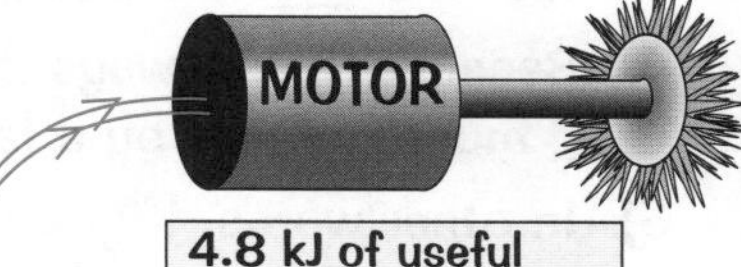

4.8 kJ of useful energy in 2 minutes

Power is Measured in Watts (or J/s)

The proper unit of power is the watt (W). One watt = 1 joule of energy transferred per second. Power means "how much energy per second", so watts are the same as "joules per second" (J/s). Don't ever say "watts per second" — it's nonsense.

Force yourself to do some work and learn this page...

Work done can sound like quite a wishy washy term, but the thing to remember is that it's just the energy transferred to something when you force it to move some distance. Like when you push a shopping trolley — you apply a force and cruise down the aisles as you bask in the glory of having transferred chemical energy to kinetic.

Kinetic Energy

Anything that's moving has kinetic energy. There's a slightly tricky formula for it, so you have to concentrate a little bit harder for this one. But hey, that's life — it can be real tough sometimes.

Kinetic Energy is Energy of Movement

You might also see kinetic energy written 'E_k'.

1) The kinetic energy (K.E.) of something is the energy it has when moving.
2) The kinetic energy of something depends on both its mass and speed.
3) The greater its mass and the faster it's going, the bigger its kinetic energy will be.
4) For example, a high-speed train, or a speedboat, will have lots of kinetic energy — but your gran doing the weekly shop on her little scooter will only have a little bit.
5) You need to know how to use the formula:

$$\text{Kinetic Energy} = \tfrac{1}{2} \times \text{mass} \times \text{speed}^2$$

K.E. / $\tfrac{1}{2} \times m \times v^2$

EXAMPLE: A car of mass 1450 kg is travelling at 28 m/s. Calculate its kinetic energy.

ANSWER: It's pretty easy. You just plug the numbers into the formula — but watch the 'v^2'!
K.E. = $\tfrac{1}{2}mv^2 = \tfrac{1}{2} \times 1450 \times 28^2$ = 568 400 J. (Joules because it's energy.)

6) If you double the mass, the K.E. doubles. If you double the speed, though, the K.E. quadruples (increases by a factor of 4) — it's because of the 'v^2' in the formula.

Stopping Distances Increase Alarmingly with Extra Speed — Mainly Because of the v^2 Bit in the K.E. Formula

1) To stop a car, the kinetic energy, $\tfrac{1}{2}mv^2$, has to be converted to heat energy at the brakes and tyres:

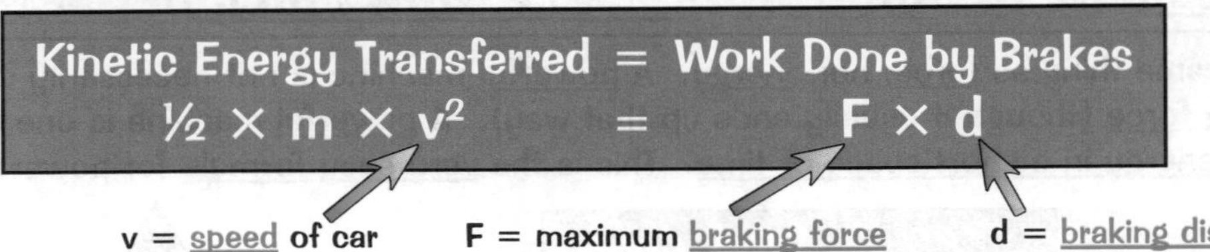

2) The braking distance (d) increases as speed squared (v^2) increases — it's a squared relationship.
3) This means if you double the speed, you double the value of v, but the v^2 means that the K.E. is then increased by a factor of four.
4) Because 'F' is always the maximum possible braking force (which can't be increased), d must increase by a factor of four to make the equation balance.
5) In other words, if you go twice as fast, the braking distance must increase by a factor of four to convert the extra K.E.
6) Increasing the speed by a factor of 3 increases the K.E. by a factor of 3^2 (= 9), so the braking distance becomes 9 times as long.
7) Doubling the mass of the object doubles the K.E. it has — which will double the braking distance. So a big heavy lorry will need more space to stop than a small car.

Look back at page 47 for more on braking distances.

Kinetic energy — just get a move on and learn it, OK...

When meteors and space shuttles enter the atmosphere, they have loads and loads of kinetic energy. Friction with the air transfers kinetic energy to heat — so much heat that most meteors burn up completely and never hit us. Space shuttles have heat shields that lose heat quickly, so they can re-enter the atmosphere without burning up.

Gravitational Potential Energy

It's a physics page... and it's got roller coasters on it. Life is good.

Gravitational Potential Energy is Energy Due to Height

Gravitational Potential Energy = mass × g × height

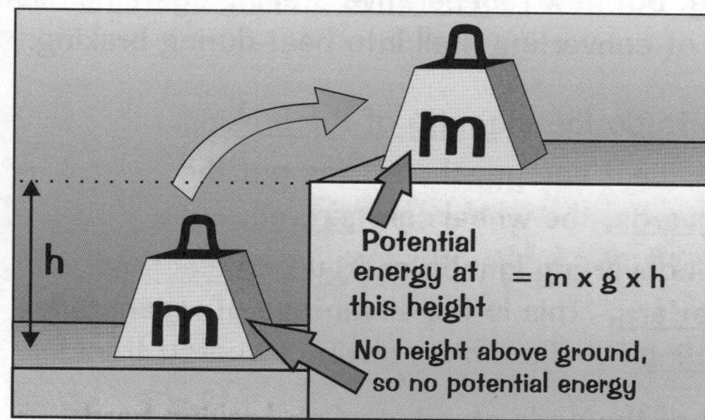

Gravitational potential energy (measured in joules) is the energy that an object has by virtue of (because of) its vertical position in a gravitational field. When an object is raised vertically, work is done against the force of gravity (it takes effort to lift it up) and the object gains gravitational potential energy. On Earth the gravitational field strength (g) is approximately 10 N/kg.

You might also see gravitational potential energy written 'E_p'.

Falling Objects Convert G.P.E. into K.E.

1) When something falls, its gravitational potential energy is converted into kinetic energy (K.E.). So the further it falls, the faster it goes.
2) In practice, some of the G.P.E. will be dissipated as heat due to air resistance, but in exam questions they'll likely say you can ignore air resistance, in which case you'll just need to remember this simple and really quite obvious formula:

K.E. gained = G.P.E. lost

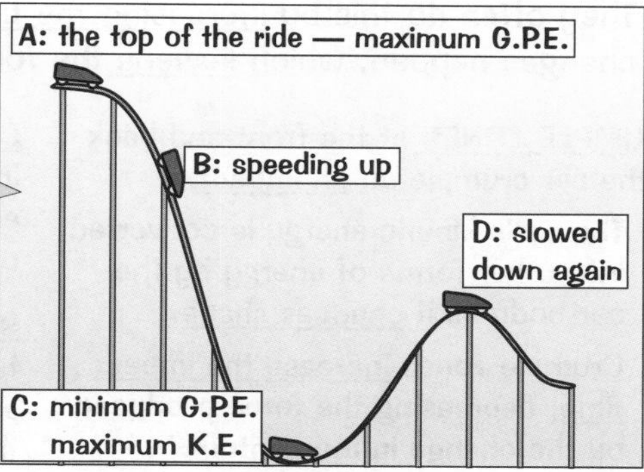

3) For example, the roller coaster to the right will lose G.P.E. and gain K.E. as it falls between points A and C.
4) If you ignore friction (between the tracks and the wheels) and air resistance, the amount of K.E. it gains will be the same as the amount of G.P.E. it loses.
5) Between C and D, it's gaining height, so some of that K.E. is converted back to G.P.E. again.

EXAMPLE: The carriage in the diagram has a mass of 500 kg and the vertical height difference between A and C is 20 m.

a) Ignoring friction and air resistance, how much K.E. is gained by the carriage in moving from A to C?

b) The roller coaster was stationary at A. Calculate its speed at C.

ANSWER: a) K.E. gained = G.P.E. lost = mass × g × height = 500 × 10 × 20 = 100 000 J

b) At C it has 100 000 J of K.E. You know that K.E. = ½mv² (see previous page), so...

$\frac{1}{2} \times m \times v^2 = 100\,000$

$v^2 = 100\,000 \div (\frac{1}{2} \times m) = 100\,000 \div (\frac{1}{2} \times 500) = 400$

$v = \sqrt{400} = 20$ m/s

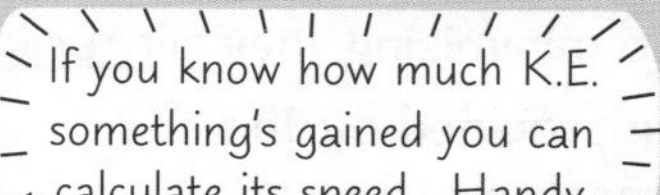

Revise roller coasters — don't let your thoughts wander off into oblivion...

Roller coasters are constantly transferring between gravitational potential and kinetic energy. In reality, energy will be lost due to friction, air resistance and even as sound. But in exams you can usually ignore these.

Car Design and Safety

A lot of the physics from the last few pages can be applied in the real world to designing safe, efficient cars.

Brakes do Work Against the Kinetic Energy of the Car

When you apply the brakes to slow down a car, work is done (see p.49).
The brakes reduce the kinetic energy of the car by transferring it into heat (and sound) energy.
In traditional braking systems that would be the end of the story, but new regenerative braking systems used in some electric or hybrid cars make use of the energy, instead of converting it all into heat during braking.

1) Regenerative brakes use the system that drives the vehicle to do the majority of the braking.
2) Rather than converting the kinetic energy of the vehicle into heat energy, the brakes put the vehicle's motor into reverse. With the motor running backwards, the wheels are slowed.
3) At the same time, the motor acts as an electric generator, converting kinetic energy into electrical energy that is stored as chemical energy in the vehicle's battery. This is the advantage of regenerative brakes — they store the energy of braking rather than wasting it. It's a nifty chain of energy transfer.

ABS (anti-lock braking system) brakes help drivers keep control of the car's steering when braking hard. They automatically pump on and off to stop the wheels locking and prevent skidding.

Cars are Designed to Convert Kinetic Energy Safely in a Crash

1) If a car crashes it will slow down very quickly — this means that a lot of kinetic energy is converted into other forms of energy in a short amount of time, which can be dangerous for the people inside.
2) In a crash, there'll be a big change in momentum (see p.48) over a very short time, so the people inside the car experience huge forces that could be fatal.
3) Cars are designed to convert the kinetic energy of the car and its passengers in a way that is safest for the car's occupants. They often do this by increasing the time over which momentum changes happen, which lessens the forces on the passengers.

airbag

seat belt

CRUMPLE ZONES at the front and back of the car crumple up on impact.

- The car's kinetic energy is converted into other forms of energy by the car body as it changes shape.
- Crumple zones increase the impact time, decreasing the force produced by the change in momentum.

SIDE IMPACT BARS are strong metal tubes fitted into car door panels. They help direct the kinetic energy of the crash away from the passengers to other areas of the car, such as the crumple zones.

SEAT BELTS stretch slightly, increasing the time taken for the wearer to stop. This reduces the forces acting in the chest. Some of the kinetic energy of the wearer is absorbed by the seat belt stretching.

AIR BAGS also slow you down more gradually and prevent you from hitting hard surfaces inside the car.

Cars Have Different Power Ratings

Sports car power = 100 kW

1) The size and design of car engines determine how powerful they are.
2) The more powerful an engine is, the more energy it transfers from its fuel every second, and so the faster its top speed can be.
3) E.g. the power output of a typical small car will be around 50 kW and a sports car will be about 100 kW (some are much higher).

Small car power = 50 kW

4) Cars are also designed to be aerodynamic. This means that they are shaped in such a way that air flows very easily and smoothly past them, so minimising their air resistance.

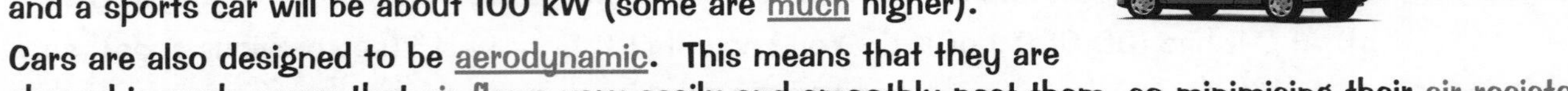

5) Cars reach their top speed when the resistive force equals the driving force provided by the engine (see p.46). So, with less air resistance to overcome, the car can reach a higher speed before this happens. Aerodynamic cars therefore have higher top speeds.

Don't let all this revising drive you crazy...

A car's safety features are tested using crash test dummies to make sure it's safe. The dummies have sensors at different places on their 'bodies' to show where a real person would be injured, and how bad the injury would be.

Forces and Elasticity

Forces aren't just important for cars and falling sheep — you can stretch things with them as well. It can sound quite tricky at first, but it's not as hard as it looks.

Work Done to an Elastic Object is Stored as Elastic Potential Energy

1) When you apply a force to an object you may cause it to stretch and change in shape.
2) Any object that can go back to its original shape after the force has been removed is an elastic object.
3) Work is done to an elastic object to change its shape. This energy is not lost but is stored by the object as elastic potential energy.
4) The elastic potential energy is then converted to kinetic energy when the force is removed and the object returns to its original shape, e.g. when a spring or an elastic band bounces back.

Elastic potential energy — useful for passing exams and scaring small children

Extension of an Elastic Object is Directly Proportional to Force...

If a spring is supported at the top and then a weight attached to the bottom, it stretches.

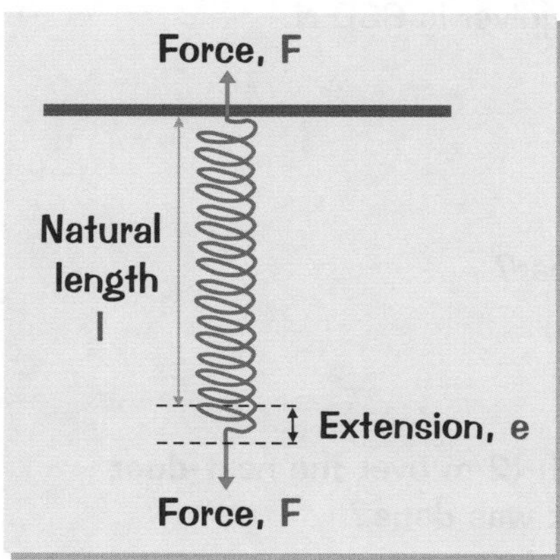

1) The extension, e, of a stretched spring (or other elastic object) is directly proportional to the load or force applied, F. The extension is measured in metres, and the force is measured in newtons.
2) This is the equation you need to learn:

$$F = k \times e$$

3) k is the spring constant. Its value depends on the material that you are stretching and it's measured in newtons per metre (N/m).

...but this Stops Working when the Force is Great Enough

There's a limit to the amount of force you can apply to an object for the extension to keep on increasing proportionally.

1) The graph shows force against extension for an elastic object.
2) For small forces, force and extension are proportional. So the first part of the graph shows a straight-line relationship between force and extension.
3) There is a maximum force that the elastic object can take and still extend proportionally. This is known as the limit of proportionality and is shown on the graph at the point marked P.
4) If you increase the force past the limit of proportionality, the material will be permanently stretched. When the force is removed, the material will be longer than at the start.

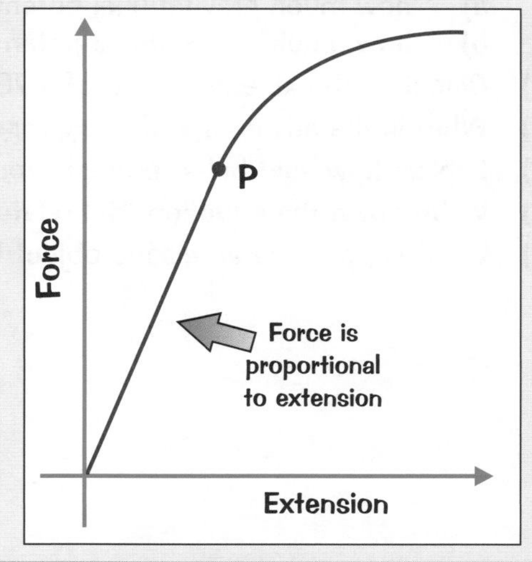

I could make a joke, but I don't want to stretch myself...

Scaring small children aside, elastic potential is really quite a useful form of energy. Think of all the things we rely on that use it — catapults, trampolines, scrunchies... Ah, elastic potential energy — thank you for enriching our lives.

Revision Summary for Section Three

Well done — you've made it to the end of another section. There are loads of bits and bobs about forces, motion and fast cars which you have to learn. The best way to find out what you know is to get stuck in to these lovely revision questions, which you're going to really enjoy (honest)...

1) What's the difference between speed and velocity?
2) What is acceleration? What are its units?
3)* Write down the formula for acceleration. What's the acceleration of a soggy pea flicked from rest to a speed of 14 m/s in 0.4 seconds?
4) Sketch a typical distance-time graph and point out all the important parts of it.
5) Explain how to calculate speed from a distance-time graph.
6) Sketch a typical velocity-time graph and point out all the important parts of it.
7) Explain how to find speed, distance and acceleration from a velocity-time graph.
8) Explain the difference between mass and weight. What units are they measured in?
9) Explain what is meant by a "resultant force".
10) If an object has zero resultant force on it, can it be moving? Can it be accelerating?
11)* Write down the formula relating resultant force and acceleration.
A resultant force of 30 N pushes a trolley of mass 4 kg. What will be its acceleration?
12)* A skydiver has a mass of 75 kg. At 80 mph, the drag force on the skydiver is 650 N.
Find the acceleration of the skydiver at 80 mph (take g = 10 N/kg).
13)* A yeti pushes a tree with a force of 120 N. What is the size of the reaction force that the Yeti feels pushing back at him?
14) What is "terminal velocity"?
15) What are the two different parts of the overall stopping distance of a car?
16) Write down the formula for momentum.
17) If the total momentum of a system before a collision is zero, what is the total momentum of the system after the collision?
18)* Write down the formula for work done. A crazy dog drags a big branch 12 m over the next-door neighbour's front lawn, pulling with a force of 535 N. How much work was done?
19)* What's the formula for kinetic energy? Find the kinetic energy of a 78 kg sheep moving at 23 m/s.
20)* A car of mass 1000 kg is travelling at a velocity of 2 m/s when a dazed and confused sheep runs out 5 m in front. If the driver immediately applies the maximum braking force of 395 N, can he avoid hitting it?
21)* A 4 kg cheese is taken 30 m up a hill before being rolled back down again. If g = 10 N/kg:
    a) how much gravitational potential energy does the cheese have at the top of the hill?
    b) how much gravitational potential energy does it have when it gets half way down?
22)* Calculate the kinetic energy of a 78 kg sheep just as she hits the floor after falling through 20 m.
23) What is the advantage of using regenerative braking systems?
24) Explain how seat belts, crumple zones, side impact bars and air bags are useful in a crash.
25) Write down the equation that relates the force on a spring and its extension.
26) What happens to an elastic object that is stretched beyond its limit of proportionality?

* Answers on page 140.

Static Electricity

Static electricity's all about charges which are not free to move. This causes them to build up in one place, and lead to sparks or shocks when they finally do move — crackling when you take a jumper off, say.

Build-up of Static is Caused by Friction

1) When two insulating materials are rubbed together, electrons are scraped off one and dumped on the other.
2) Electrons are negatively charged.
3) So this leaves a positive static charge on one (electrons scrapped off) and a negative static charge on the other (gained electrons).
4) Which way the electrons are transferred depends on the two materials involved.
5) The classic examples are polythene and acetate rods being rubbed with a cloth duster, as shown in the diagrams.

With the polythene rod, electrons move from the duster to the rod.

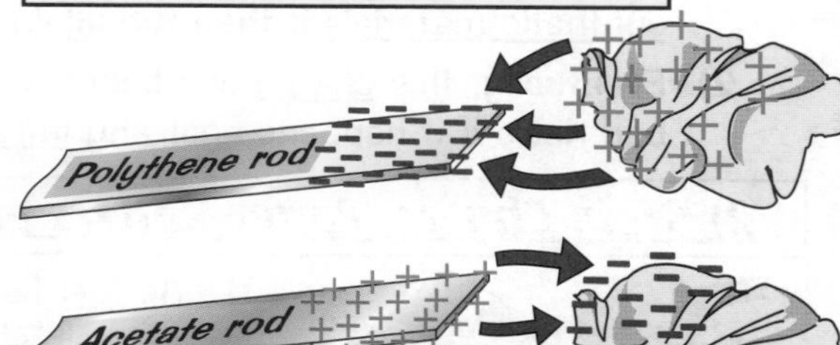

With the acetate rod, electrons move from the rod to the duster.

Only Electrons Move — Never the Positive Charges

When electrons are removed from particles the particles are left positively charged — these charged particles are called ions (see p.35).

Both +ve and –ve electrostatic charges are only ever produced by the movement of electrons — the negatively charged particles. The positive charges definitely do not move. A positive static charge is always caused by electrons moving away elsewhere, as shown above. Don't forget!

Like Charges Repel, Opposite Charges Attract

Two things with opposite electric charges are attracted to each other.
Two things with the same electric charge will repel each other.

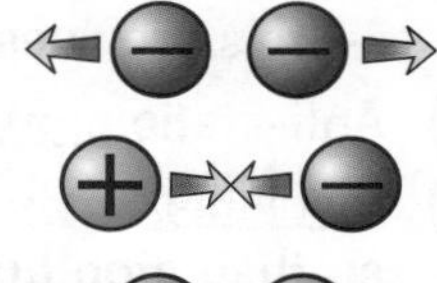

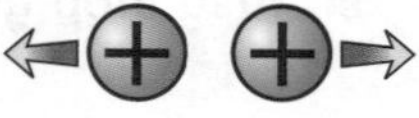

When you rub two insulating materials together a whole load of electrons get dumped together on one of the insulators, which becomes negatively charged. They try to repel each other, but can't move apart because their positions are fixed. The patch of charge that results is called static electricity because it can't move.

Static Electricity can be a Little Joker

Static electricity is responsible for some of life's little annoyances...

1) Attracting Dust

Dust particles are really tiny and lightweight and are easily attracted to anything that's charged. Unfortunately, many objects around the house are made of insulating materials (e.g. glass, wood, plastic) that get easily charged and attract the dust particles — this makes cleaning a nightmare. (Have a look at how dusty your TV screen is.)

2) Clinging Clothes and Crackles

When synthetic clothes are dragged over each other (like in a tumble drier) or over your head, electrons get scraped off, leaving static charges on both parts, and that leads to the inevitable — attraction (they stick together and cling to you) and little sparks or shocks as the charges rearrange themselves.

3) Bad Hair Days

Static builds up on your hair, giving each strand the same charge — so they repel each other.

Static caravans — where electrons go on holiday...

Static electricity's great fun. You must have tried it — rubbing a balloon against your jumper and trying to get it to stick to the ceiling. It really works... well, sometimes, and if at first you don't succeed, try, try again...

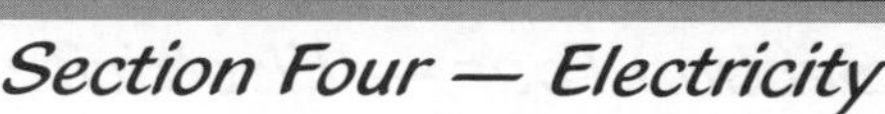

Dangers and Uses of Static Electricity

Static Electricity Can be Dangerous...

1) A Lot of Charge Can Build Up on Clothes

1) A large amount of static charge can build up on clothes made out of synthetic materials if they rub against other synthetic fabrics (see p.55).
2) Eventually, this charge can become large enough to make a spark — which is really bad news if it happens near any inflammable gases or fuel fumes... KABOOM!

2) Grain Chutes, Paper Rollers and the Fuel Filling Nightmare

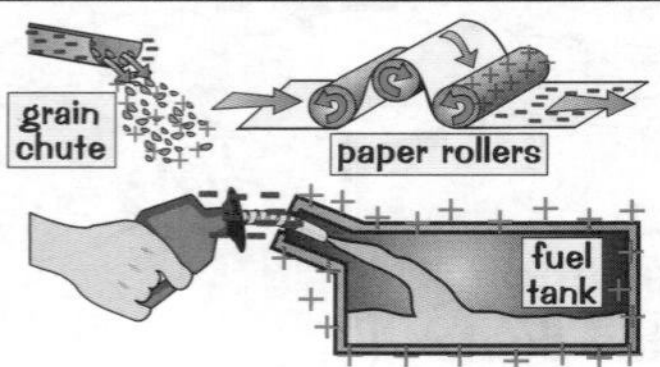

1) As fuel flows out of a filler pipe, or paper drags over rollers, or grain shoots out of pipes, then static can build up.
2) This can easily lead to a spark and might cause an explosion in dusty or fumey places — like when filling up a car with fuel at a petrol station.
3) All these problems with sparks can be solved by earthing charged objects.

Objects Can be Earthed or Insulated to Prevent Sparks

1) Dangerous sparks can be prevented by connecting a charged object to the ground using a conductor (e.g. a copper wire) — this is called earthing and it provides an easy route for the static charges to travel into the ground. This means no charge can build up to give you a shock or make a spark.
2) Static charges are a big problem in places where sparks could ignite inflammable gases, or where there are high concentrations of oxygen (e.g. in a hospital operating theatre).
3) Fuel tankers must be earthed to prevent any sparks that might cause the fuel to explode — refuelling aircraft are bonded to their fuel tankers using an earthing cable to prevent sparks.
4) Anti-static sprays and liquids work by making the surface of a charged object conductive — this provides an easy path for the charges to move away and not cause a problem.
5) Anti-static cloths are conductive, so they can carry charge away from objects they're used to wipe.
6) Insulating mats and shoes with insulating soles prevent static electricity from moving through them, so they stop you from getting a shock.

...But it can also be Pretty Useful

1) Bikes and cars are painted using electrostatic paint sprayers. The spray gun is charged, which charges up the small drops of paint. Each paint drop repels all the others, since they've all got the same charge, so you get a very fine spray. The object to be painted is given an opposite charge to the gun and attracts the fine spray of paint. This method gives an even coat, hardly any paint is wasted and parts of the object pointing away from the spray gun still receive paint too — there are no paint shadows.
2) Dust precipitators use static electricity to clean up emissions from factories and power stations. Dust particles become negatively charged as they pass through a charged wire grid in the chimney. The negatively charged dust particles then stick to earthed metal plates and eventually fall to the bottom of the chimney where they can be removed.

Chimney
Earthed metal plates
Negatively charged grid

3) An electric shock from a defibrillator can restart a stopped heart. The defibrillator consists of two paddles connected to a power supply which are placed firmly on the patient's chest. The defibrillator operator holds insulated handles — so only the patient gets a shock. The charge passes through the paddles to the patient to make the heart contract.

Static electricity — it's really shocking stuff...

Lightning is an extreme case of a static electricity spark. It always chooses the easiest path between the sky and the ground — that's the nearest, tallest thing. That's why it's never a good idea to fly a kite in a thunderstorm...

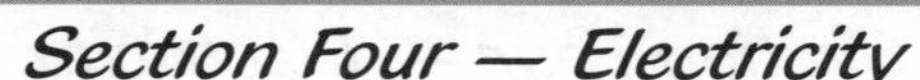

Current and Potential Difference

1) Current is the flow of electric charge round the circuit. Current will only flow through a component if there is a potential difference across that component. Unit: ampere, A.

2) Potential Difference is the driving force that pushes the current round. Unit: volt, V.

3) Resistance is anything in the circuit which slows the flow down. Unit: ohm, Ω.

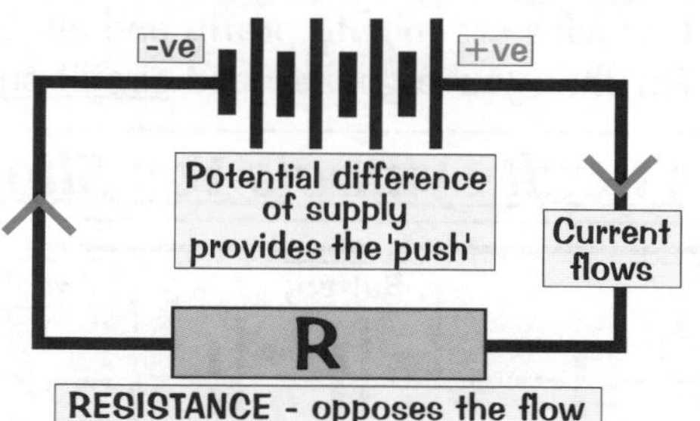

The greater the resistance across a component, the smaller the current that flows (for a given potential difference across the component).

Voltage and potential difference are the same thing.

Total Charge Through a Circuit Depends on Current and Time

1) Current is the rate of flow of charge. When current (I) flows past a point in a circuit for a length of time (t) then the charge (Q) that has passed is given by this formula:

2) Current is measured in amperes (A), charge is measured in coulombs (C), time is measured in seconds (s).

$$\text{Current} = \frac{\text{Charge}}{\text{Time}} \qquad I = \frac{Q}{t}$$

3) More charge passes around the circuit when a bigger current flows.

EXAMPLE: A battery charger passes a current of 2.5 A through a cell over a period of 4 hours. How much charge does the charger transfer to the cell altogether?

ANSWER: $Q = I \times t = 2.5 \times (4 \times 60 \times 60) = 36\ 000$ C (36 kC).

Potential Difference (P. D.) is the Work Done Per Unit Charge

1) The potential difference (or voltage) is the work done (the energy transferred, measured in joules, J) per coulomb of charge that passes between two points in an electrical circuit. It's given by this formula:

2) So, the potential difference across an electrical component is the amount of energy that is transferred by that electrical component (e.g. to light and heat energy by a bulb) per unit of charge.

$$\text{P.D.} = \frac{\text{Work done}}{\text{Charge}}$$

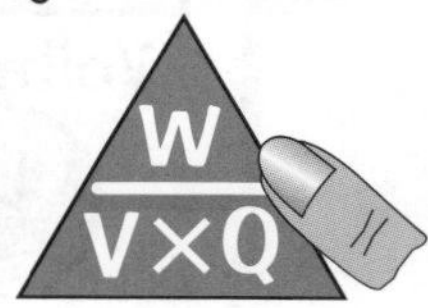

A Voltmeter Measures Potential Difference Between Two Points

1) A battery transfers energy to the charge as it passes — that's the "push" that moves the charge round the circuit.

2) Components transfer energy away from the charge as it passes — e.g. to use as light in a lamp or sound in a buzzer.

3) The voltage of a battery shows how much work the battery will do to charge that passes through it (how big a "push" it gives it).

4) A voltmeter is used to measure the potential difference between two points.

5) A voltmeter must be placed in parallel (see p.62) with a component so it can compare the energy the charge has before and after passing through the component (as in the diagram).

6 V
0 V | 6 V
The battery transfers energy to the charge as it passes.
direction that current is moving
0 V | 6 V
6 V
The lamp transfers the same amount of energy from the charge as it passes (and converts it to light and heat).

I think it's about time you took charge...

An interesting fact — voltage is named after Count Alessandro Volta, an Italian physicist. I heard once that potential difference was named after his cousin — Baron Potentialo Differenché. I'm not so sure if it's true...

Circuits — The Basics

Formulas are mighty pretty and all, but you might have to design some electrical circuits as well one day. For that you're gonna need circuit symbols. Well, would you look at that... they're on this page.

Circuit Symbols You Should Know

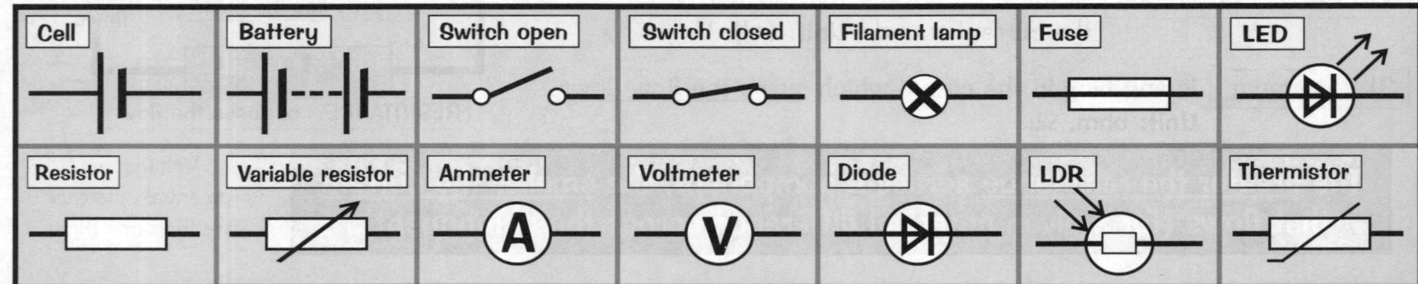

The Standard Test Circuit

This is the circuit you use if you want to know the resistance of a component. You find the resistance by measuring the current through and the potential difference across the component. It is absolutely the most bog standard circuit you could know.

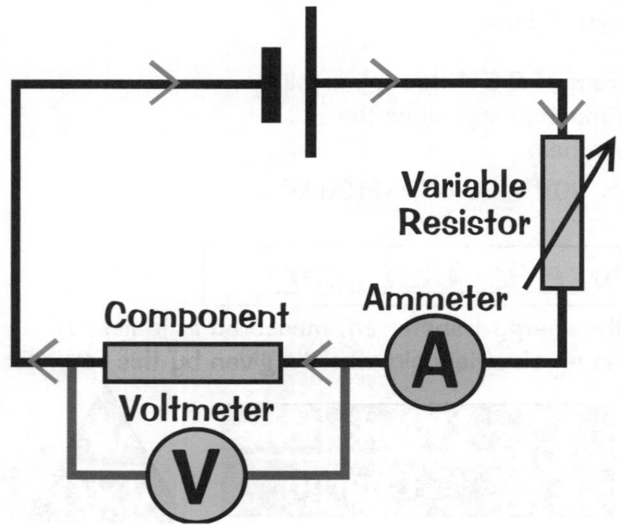

The Ammeter

1) Measures the current (in amps) flowing through the component.
2) Must be placed in series (see p.61).
3) Can be put anywhere in series in the main circuit, but never in parallel like the voltmeter.

The Voltmeter

1) Measures the potential difference (in volts) across the component.
2) Must be placed in parallel (see p.62) around the component under test — NOT around the variable resistor or the battery!

Five Important Points

1) This very basic circuit is used for testing components, and for getting V-I graphs from them (see next page).
2) The component, the ammeter and the variable resistor are all in series, which means they can be put in any order in the main circuit. The voltmeter, on the other hand, can only be placed in parallel around the component under test, as shown. Anywhere else is a definite no-no.
3) As you vary the variable resistor it alters the current flowing through the circuit.
4) This allows you to take several pairs of readings from the ammeter and voltmeter.
5) You can then plot these values for current and voltage on a V-I graph and find the resistance.

Measure gymnastics — use a vaultmeter...

The funny thing is — the electrons in circuits actually move from –ve to +ve... but scientists always think of current as flowing from +ve to –ve. Basically it's just because that's how the early physicists thought of it (before they found out about the electrons), and now it's become convention.

Resistance and $V = I \times R$

With your current and your potential difference measured, you can now make some sweet graphs...

Three Hideously Important Potential Difference-Current Graphs

V-I graphs show how the current varies as you change the potential difference (P.D.). Here are three examples:

Different Resistors

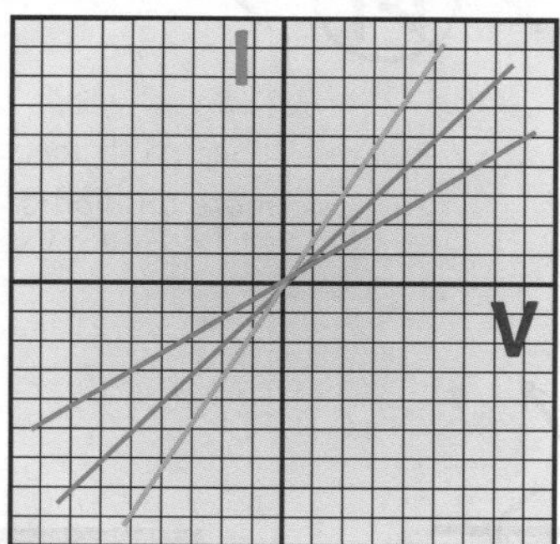

The current through a resistor (at constant temperature) is directly proportional to P.D. Different resistors have different resistances, hence the different slopes.

Filament Lamp

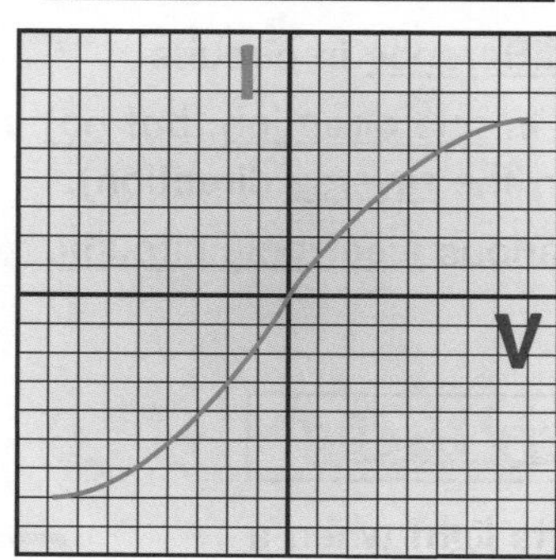

As the temperature of the filament increases, the resistance increases, hence the curve.

Diode

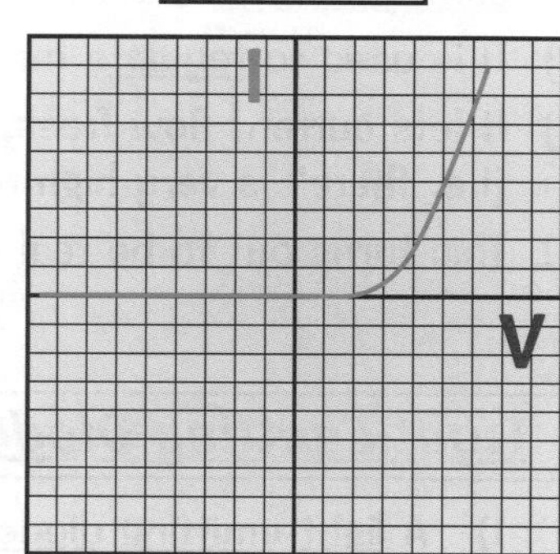

Current will only flow through a diode in one direction, as shown. The diode has very high resistance in the opposite direction.

Resistance Increases with Temperature

1) When an electrical charge flows through a resistor, some of the electrical energy is transferred to heat energy and the resistor gets hot.
2) This heat energy causes the ions in the conductor to vibrate more. With the ions jiggling around it's more difficult for the charge-carrying electrons to get through the resistor — the current can't flow as easily and the resistance increases.
3) For most resistors there is a limit to the amount of current that can flow. More current means an increase in temperature, which means an increase in resistance, which means the current decreases again.
4) This is why the graph for the filament lamp levels off at high currents.

Resistance, Potential Difference and Current: $V = I \times R$

Potential Difference = Current × Resistance

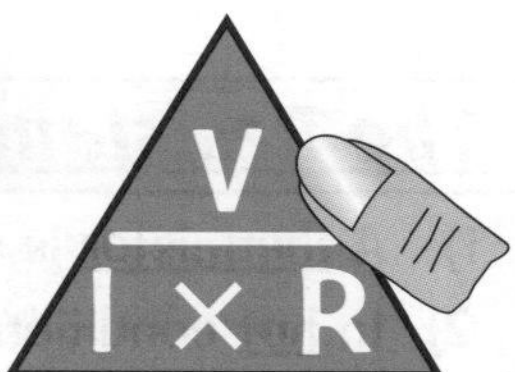

For the straight-line graphs above, the resistance of the component is steady and is equal to the inverse of the gradient of the line, or "1/gradient". In other words, the steeper the graph the lower the resistance.

If the graph curves, it means the resistance is changing. In that case R can be found for any point by taking the pair of values (V, I) from the graph and sticking them in the formula $R = V/I$. Easy.

EXAMPLE: Voltmeter V reads 6 V and resistor R is 4 Ω. What is the current through Ammeter A?

ANSWER: Use the formula triangle for $V = I \times R$. We need to find I, so the version we need is $I = V/R$. The answer is then: $I = 6 \div 4 = 1.5$ A.

In the end you'll have to learn this — resistance is futile...

You might need to interpret some potential difference-current graphs for your exam. Remember — the steeper the slope, the lower the resistance. And you need to know that formula inside out, back to front, upside down and in Swahili. It's the most important equation in electrics, bar none. (P.S. I might let you off the Swahili.)

Circuit Devices

You might consider yourself a bit of an expert in circuit components — you're enlightened about bulbs, you're switched on to switches... But take a look at these ones as well — they're a little bit trickier.

Current Only Flows in One Direction through a Diode

1) A diode is a special device made from semiconductor material such as silicon.
2) It is used to regulate the potential difference in circuits.
3) It lets current flow freely through it in one direction, but not in the other (i.e. there's a very high resistance in the reverse direction).
4) This turns out to be real useful in various electronic circuits.

See p.135 for more on diodes.

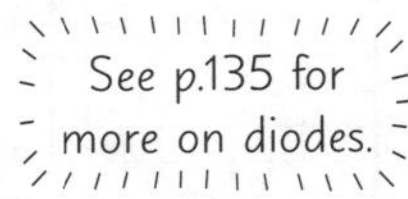

Light-Emitting Diodes are Very Useful

1) A light-emitting diode (LED) emits light when a current flows through it in the forward direction.
2) LEDs are being used more and more as lighting, as they use a much smaller current than other forms of lighting.
3) LEDs indicate the presence of current in a circuit. They're often used in appliances (e.g. TVs) to show that they are switched on.
4) They're also used for the numbers on digital clocks, in traffic lights and in remote controls.

See p.134 for more on LEDs.

A Light-Dependent Resistor or "LDR" to You

1) An LDR is a resistor that is dependent on the intensity of light. Simple really.
2) In bright light, the resistance falls.
3) In darkness, the resistance is highest.
4) They have lots of applications including automatic night lights, outdoor lighting and burglar detectors.

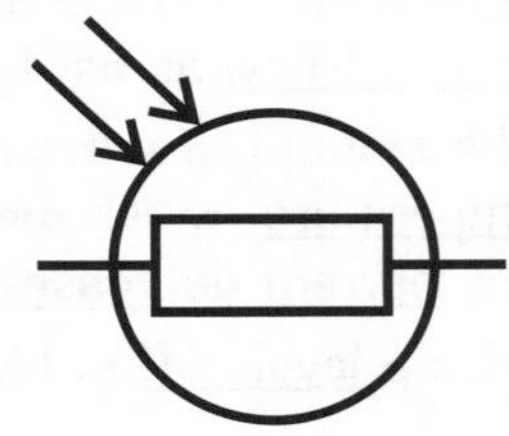

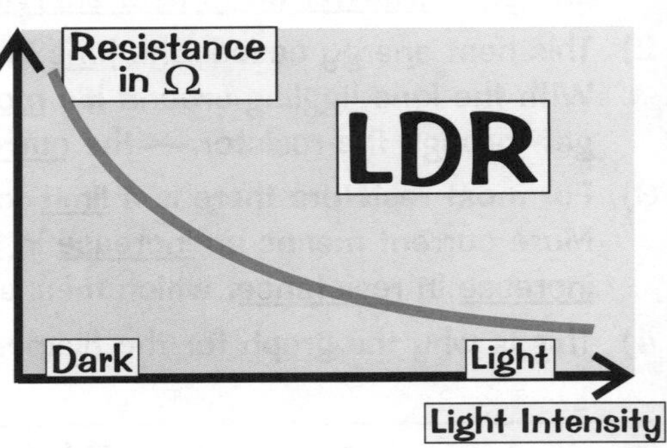

The Resistance of a Thermistor Decreases as Temperature Increases

1) A thermistor is a temperature dependent resistor.
2) In hot conditions, the resistance drops.
3) In cool conditions, the resistance goes up.
4) Thermistors make useful temperature detectors (see page 131), e.g. car engine temperature sensors and electronic thermostats.

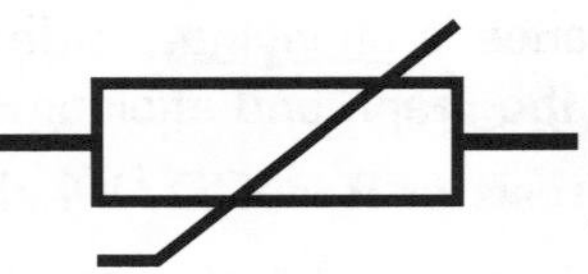

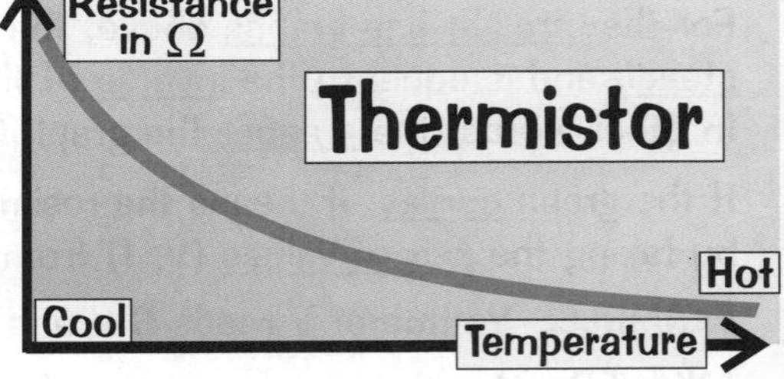

LDRs — Light-Dependent Rabbits...

LDRs are good triggers in security systems, because they can detect when the light intensity changes. So if a robber walks in front of a beam of light pointed at the LDR, the resistance shoots up and an alarm goes off.

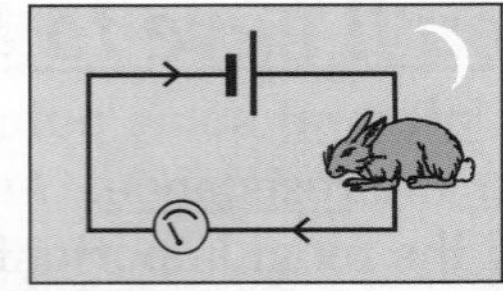

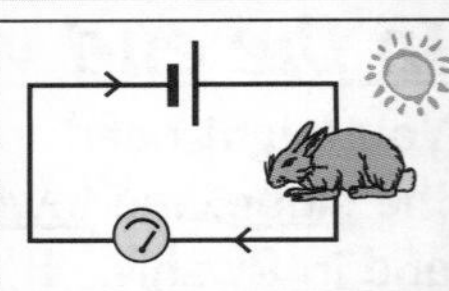

Series Circuits

You need to be able to tell the difference between series and parallel circuits just by looking at them.

Series Circuits — Everything in a Line

In series circuits, the different components are connected in a line, end to end, between the +ve and –ve of the power supply (except for voltmeters, which are always connected in parallel, but they don't count).

Potential Difference is Shared:

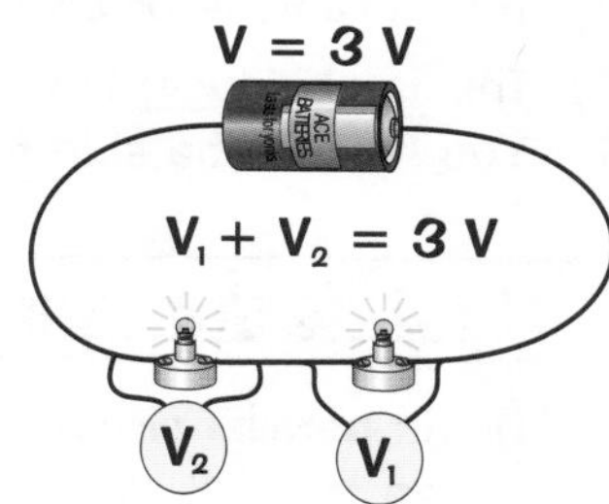

1) In series circuits, the total potential difference (P.D.) of the supply is shared between the various components. So the P.D.s round a series circuit always add up to equal the P.D. across the battery: $V = V_1 + V_2$
2) This is because the total work done on the charge by the battery must equal the total work done by the charge on the components.

Current is the Same Everywhere:

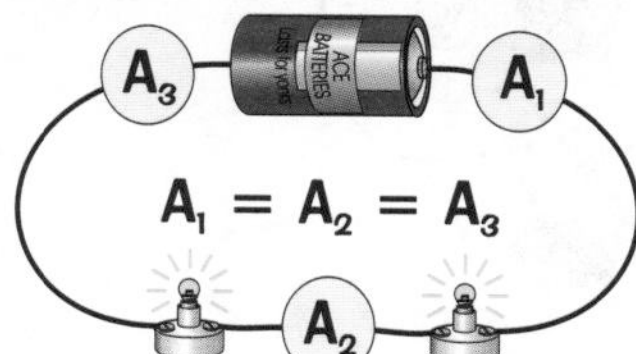

1) In series circuits the same current flows through all parts of the circuit: $A_1 = A_2 = A_3$
2) The size of the current is determined by the total P.D. of the cells and the total resistance of the circuit: i.e. I = V/R. This means all the components get the same current.

Resistance Adds Up:

1) In series circuits, the total resistance is just the sum of the individual resistances: $R = R_1 + R_2 + R_3$

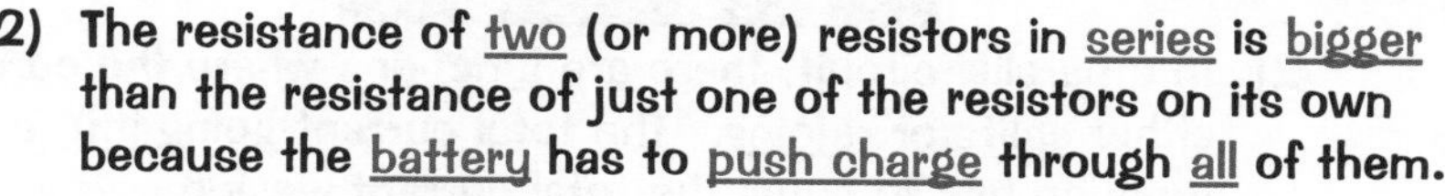

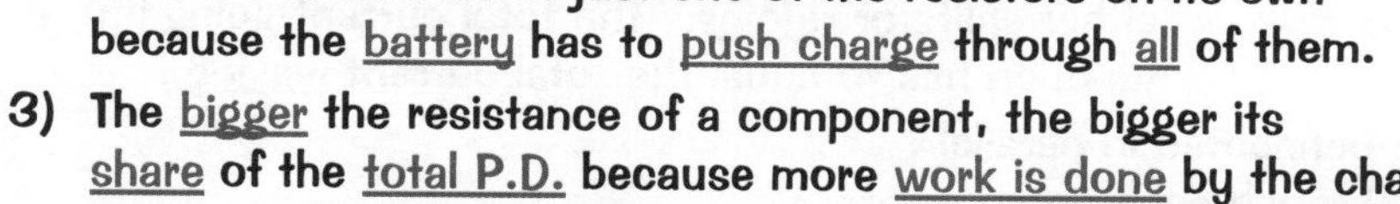

2) The resistance of two (or more) resistors in series is bigger than the resistance of just one of the resistors on its own because the battery has to push charge through all of them.
3) The bigger the resistance of a component, the bigger its share of the total P.D. because more work is done by the charge when moving through a large resistance, than through a small one.

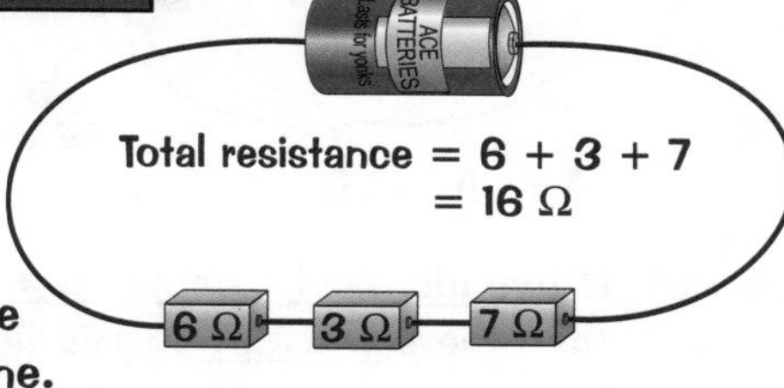

4) If the resistance of one component changes (e.g. if it's a variable resistor, light-dependent resistor or thermistor) then the potential difference across all the components will change too.

Cell Voltages Add Up:

1) If you connect several cells in series, all the same way (+ to –) you get a bigger total voltage — because each charge in the circuit passes though all the cells and gets a 'push' from each cell in turn.
2) So two 1.5 V cells in series would supply 3 V in total.
3) Cell voltages don't add up like that for cells connected in parallel. Each charge only goes through one cell.

12 V
12 V
Total = 24 V

12 V
12 V
Total = 12 V

Cell Current Doesn't Add Up:

1) Adding cells in series doesn't increase the current in a circuit. The maximum current in the circuit will just be the same as if you had one cell in the circuit.
2) Cells connected in parallel increase the total current in the circuit. However, the current through each cell is less than in the rest of the circuit because they join together to make the total current.

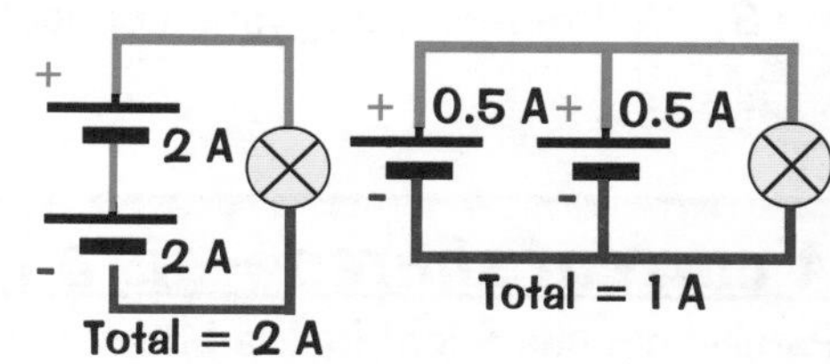

Parallel Circuits

Parallel circuits are much more sensible than series circuits and so they're much more common in real life. All the electrics in your house will be wired in parallel circuits.

Parallel Circuits — Independence and Isolation

1) In parallel circuits, each component is separately connected to the +ve and –ve of the supply.
2) If you remove or disconnect one of them, it will hardly affect the others at all.
3) This is obviously how most things must be connected, for example in cars and in household electrics. You have to be able to switch everything on and off separately.

1) P.D. is the Same Across All Components:

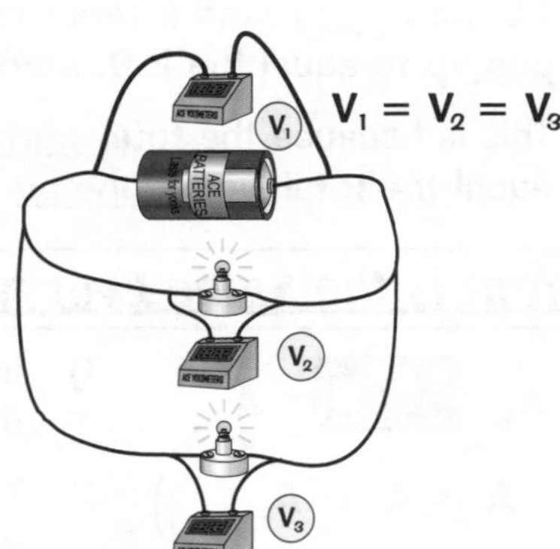

1) In parallel circuits all components get the full source P.D., so the voltage is the same across all components:

$$V_1 = V_2 = V_3$$

2) This means that identical bulbs connected in parallel will all be at the same brightness.

2) Current is Shared Between Branches:

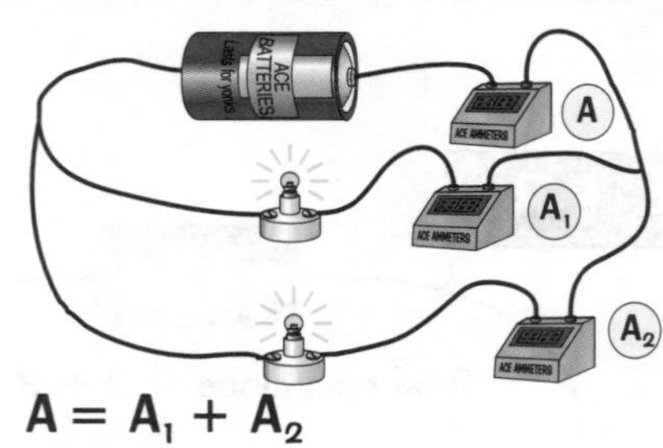

$A = A_1 + A_2$

1) In parallel circuits the total current flowing around the circuit is equal to the total of all the currents through the separate components.

$$A = A_1 + A_2 + \dots$$

2) In a parallel circuit, there are junctions where the current either splits or rejoins. The total current going into a junction has to equal the total current leaving.
3) If two identical components are connected in parallel then the same current will flow through each component.

3) Resistance Is Tricky:

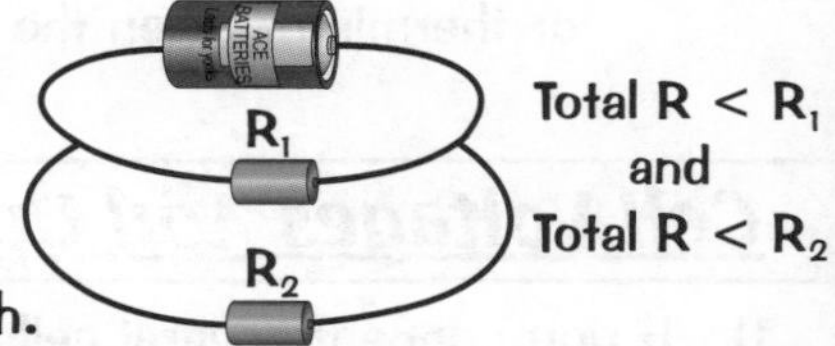

1) The total resistance of a parallel circuit is tricky to work out, but it's always less than that of the branch with the smallest resistance.
2) The resistance is lower because the charge has more than one branch to take — only some of the charge will flow along each branch.
3) A circuit with two resistors in parallel will have a lower resistance than a circuit with either of the resistors by themselves — which means the parallel circuit will have a higher current.

Voltmeters and Ammeters Are Exceptions to the Rule:

1) Ammeters and voltmeters are exceptions to the series and parallel rules.
2) Ammeters are always connected in series even in a parallel circuit.
3) Voltmeters are always connected in parallel with a component even in a series circuit.

A current shared — is a current halved...

Parallel circuits might look a bit scarier than series ones, but they're much more useful. Remember: each branch has the same voltage across it, and the total current is equal to the sum of the currents through each of the branches.

Mains Electricity

Electric current is the movement of charge carriers. To transfer energy, it doesn't matter which way the charge carriers are going. That's why an alternating current works. Read on to find out more...

Mains Supply is AC, Battery Supply is DC

1) The UK mains supply is approximately 230 volts.
2) It is an AC supply (alternating current), which means the current is constantly changing direction.
3) The frequency of the AC mains supply is 50 cycles per second or 50 Hz (hertz).
4) By contrast, cells and batteries supply direct current (DC). This just means that the current always keeps flowing in the same direction.

Electricity Supplies Can Be Shown on an Oscilloscope Screen

1) A cathode ray oscilloscope (CRO) is basically a snazzy voltmeter.
2) If you plug an AC supply into an oscilloscope, you get a 'trace' on the screen that shows how the voltage of the supply changes with time. The trace goes up and down in a regular pattern — some of the time it's positive and some of the time it's negative.
3) If you plug in a DC supply, the trace you get is just a straight line.
4) The vertical height of the AC trace at any point shows the input voltage at that point. By measuring the height of the trace you can find the potential difference of the AC supply.
5) For DC it's a lot simpler — the voltage is just the distance from the straight line trace to the centre line.

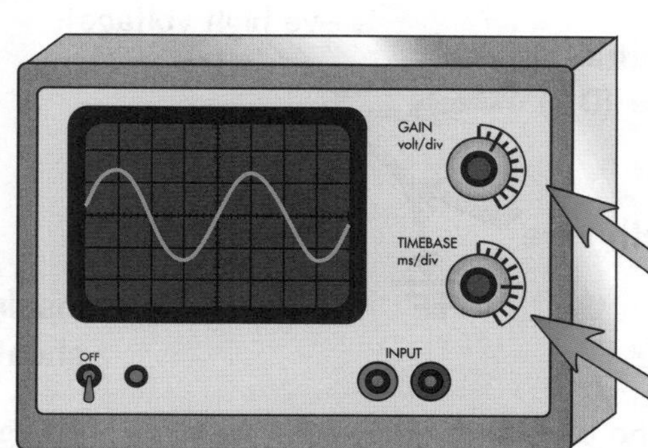

The GAIN dial controls how many volts each centimetre division represents on the vertical axis.

The TIMEBASE dial controls how many milliseconds (1 ms = 0.001 s) each division represents on the horizontal axis.

Learn How to Read an Oscilloscope Trace

DC supply

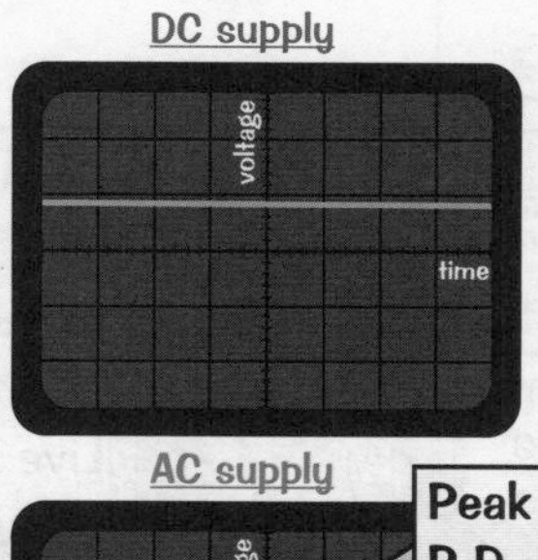

A DC source is always at the same voltage, so you get a straight line.

AC supply

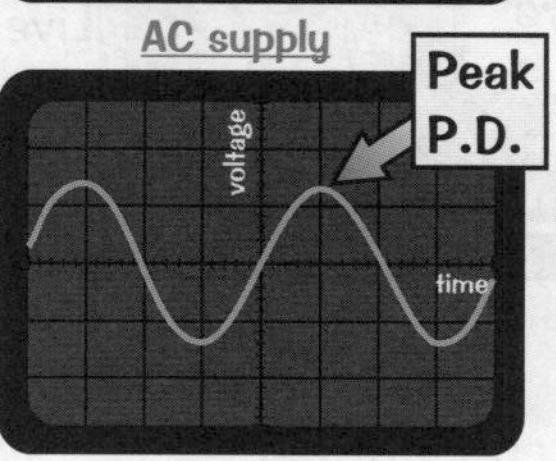

An AC source gives a regularly repeating wave. From that, you can work out the period and the frequency of the supply.

You work out the frequency using:

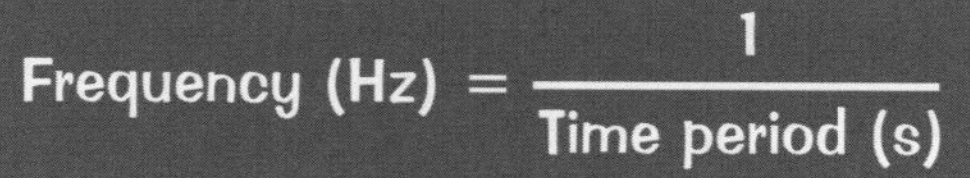

$$\text{Frequency (Hz)} = \frac{1}{\text{Time period (s)}}$$

EXAMPLE: The trace below comes from an oscilloscope with the timebase set to 5 ms/div. Find: a) the time period, and b) the frequency of the AC supply.

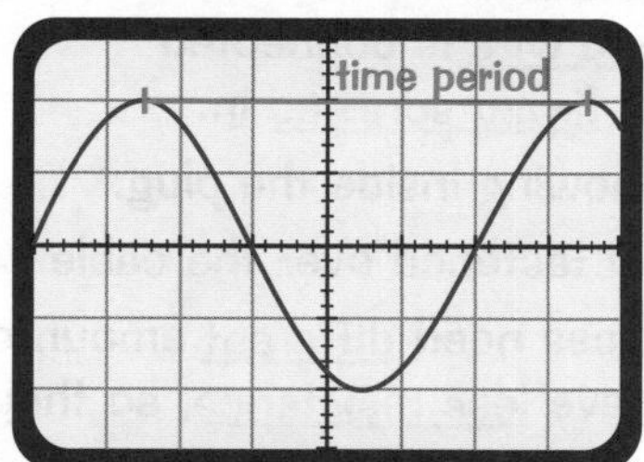

Time period = the time to complete one cycle. 1 ms = 0.001 s.

ANSWER: a) To find the time period, measure the horizontal distance between two peaks. The time period of the signal is 6 divisions. Multiply by the timebase:
Time period = 5 ms × 6 = 0.03 s

b) Using the frequency formula on the left:
Frequency = 1/0.03 = 33 Hz

I wish my bank account had a gain dial...

Because mains power is AC, its current can be increased or decreased using a device called a transformer (see pages 85-86). The lower the current in power transmission lines, the less energy is wasted as heat.

Electricity in the Home

Now then, did you know... electricity is dangerous. It can kill you. Well just watch out for it, that's all.

Hazards in the Home — Eliminate Them Before They Eliminate You

Sometimes examiners like to show you a picture of domestic bliss but with various electrical hazards in the picture such as kids shoving their fingers into sockets and stuff like that, and they'll ask you to list all the hazards. This should be mostly common sense, but it won't half help if you already know some of the likely hazards. Here are 9 examples:

1) Long cables.
2) Frayed cables.
3) Cables in contact with something hot or wet.
4) Water near sockets.
5) Shoving things into sockets.
6) Damaged plugs.
7) Too many plugs into one socket.
8) Lighting sockets without bulbs in.
9) Appliances without their covers on.

Most Cables Have Three Separate Wires

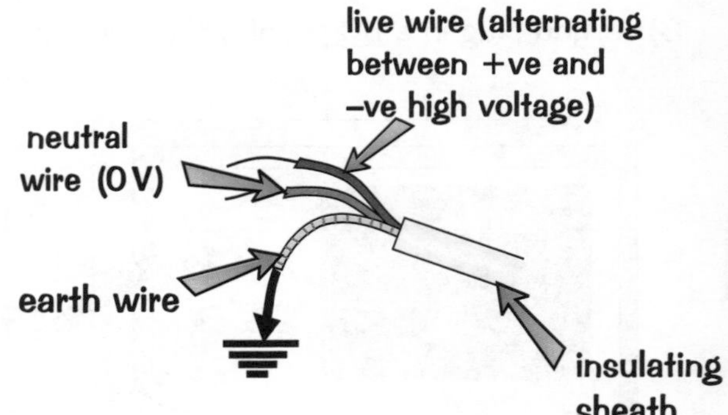

1) Most electrical appliances are connected to the mains supply by three-core cables. This means that they have three wires inside them, each with a core of copper and a coloured plastic coating.
2) The brown LIVE WIRE in a mains supply alternates between a HIGH +VE AND –VE VOLTAGE.
3) The blue NEUTRAL WIRE is always at 0 V. Electricity normally flows in and out through the live and neutral wires only.
4) The green and yellow EARTH WIRE is for protecting the wiring, and for safety — it works together with a fuse to prevent fire and shocks. It is attached to the metal casing of the plug and carries the electricity to earth (and away from you) should something go wrong and the live or neutral wires touch the metal case.

Three-Pin Plugs and Cables — Learn the Safety Features

Get the Wiring Right

1) The right coloured wire is connected to each pin, and firmly screwed in.
2) No bare wires showing inside the plug.
3) Cable grip tightly fastened over the cable outer layer.
4) Different appliances need different amounts of electrical energy. Thicker cables have less resistance, so they carry more current.

Rubber or plastic case
E
Earth Wire Green/Yellow
Fuse
Neutral Wire Blue
N
L
Live Wire Brown
Cable grip
Brass Pins

Plug Features

1) The metal parts are made of copper or brass because these are very good conductors.
2) The case, cable grip and cable insulation are made of rubber or plastic because they're really good insulators, and flexible too.
3) This all keeps the electricity flowing where it should.

CGP books are ACE — well, I had to get a plug in somewhere...

Pure water doesn't conduct electricity, but water (usually) has mineral salts dissolved in it. These carry the charge around really well, making it a very good conductor. So don't blow dry your hair in the bath, OK?

Fuses and Earthing

Questions about fuses cover a whole barrel of fun — electrical current, resistance, energy transfers and electrical safety... Read on, read on.

Earthing and Fuses Prevent Electrical Overloads

The earth wire and fuse (or circuit breaker) are included in electrical appliances for safety and work together like this:

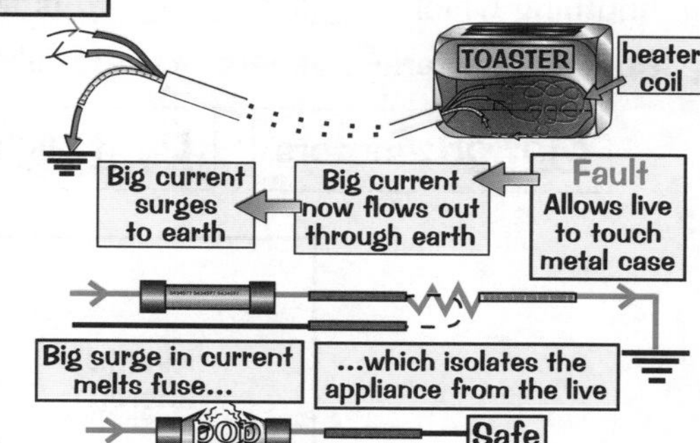

1) If a fault develops in which the live wire somehow touches the metal case, then because the case is earthed, too great a current flows in through the live wire, through the case and out down the earth wire.
2) This surge in current melts the fuse (or trips the circuit breaker in the live wire) when the amount of current is greater than the fuse rating. This cuts off the live supply and breaks the circuit.

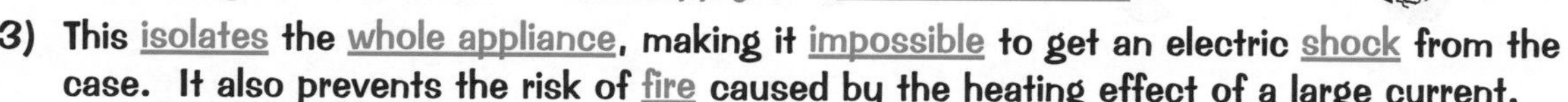

3) This isolates the whole appliance, making it impossible to get an electric shock from the case. It also prevents the risk of fire caused by the heating effect of a large current.
4) As well as people, fuses and earthing are there to protect the circuits and wiring in your appliances from getting fried if there is a current surge.
5) Fuses should be rated as near as possible but just higher than the normal operating current.
6) The larger the current, the thicker the cable you need to carry it. That's why the fuse rating needed for cables usually increases with cable thickness.

Insulating Materials Make Appliances "Double Insulated"

All appliances with metal cases are usually "earthed" to reduce the danger of electric shock. "Earthing" just means the case must be attached to an earth wire. An earthed conductor can never become live. If the appliance has a plastic casing and no metal parts showing then it's said to be double insulated.

Anything with double insulation like that doesn't need an earth wire — just a live and neutral. Cables that only carry the live and neutral wires are known as two-core cables.

Circuit Breakers Have Some Advantages Over Fuses

1) Circuit breakers are an electrical safety device used in some circuits. Like fuses, they protect the circuit from damage if too much current flows.
2) When circuit breakers detect a surge in current in a circuit, they break the circuit by opening a switch.
3) A circuit breaker (and the circuit they're in) can easily be reset by flicking a switch on the device. This makes them more convenient than fuses — which have to be replaced once they've melted.
4) They are, however, a lot more expensive to buy than fuses.
5) One type of circuit breaker used instead of a fuse and an earth wire is a Residual Current Circuit Breakers (RCCBs):
   a) Normally exactly the same current flows through the live and neutral wires. If somebody touches the live wire, a small but deadly current will flow through them to the earth. This means the neutral wire carries less current than the live wire. The RCCB detects this difference in current and quickly cuts off the power by opening a switch.
   b) They also operate much faster than fuses — they break the circuit as soon as there is a current surge — no time is wasted waiting for the current to melt a fuse. This makes them safer.
   c) RCCBs even work for small current changes that might not be large enough to melt a fuse. Since even small current changes could be fatal, this means RCCBs are more effective at protecting against electrocution.

Why are earth wires green and yellow — when mud is brown..?

All these safety precautions mean it's pretty difficult to get electrocuted on modern appliances. But that's only so long as they are in good condition and you're not doing something really stupid. Watch out for frayed wires, don't overload plugs, and for goodness sake don't use a knife to get toast out of a toaster when it is switched on.

Energy and Power in Circuits

Electricity is just another form of energy — which means that it is always conserved.

Energy is Transferred from Cells and Other Sources

Anything which supplies electricity is also supplying energy.

So cells, batteries, generators, etc. all transfer energy to components in the circuit:

Motion: motors | Light: light bulbs | Heat: Hair dryers/kettles | Sound: speakers

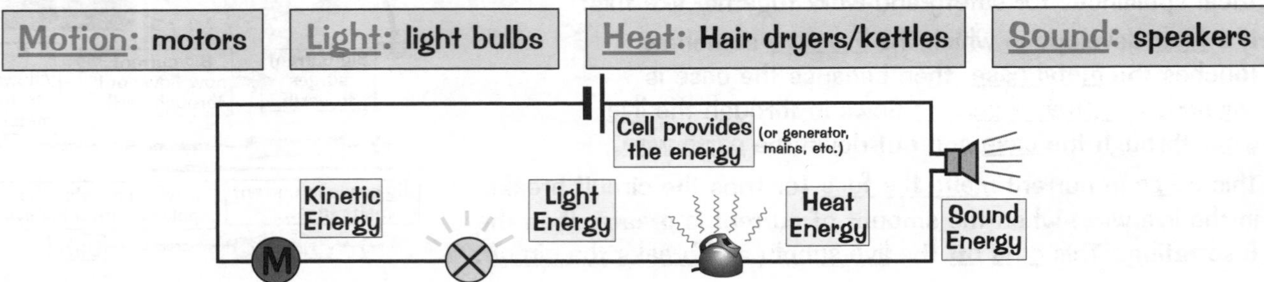

All Resistors Produce Heat When a Current Flows Through Them

1) Whenever a current flows through anything with electrical resistance (which is pretty much everything) then electrical energy is converted into heat energy.
2) The more current that flows, the more heat is produced.
3) A bigger voltage means more heating because it pushes more current through.
4) Filament bulbs work by passing a current through a very thin wire, heating it up so much that it glows. Rather obviously, they waste a lot of energy as heat.

If an Appliance is Efficient it Wastes Less Energy

All this energy wasted as heat can get a little depressing — but there is a solution.

1) When you buy electrical appliances you can choose to buy ones that are more energy efficient.
2) These appliances transfer more of their total electrical energy output to useful energy.
3) For example, less energy is wasted as heat in power-saving lamps such as compact fluorescent lamps (CFLs) and light emitting diodes (p.60) than in ordinary filament bulbs.
4) Unfortunately, they do cost more to buy, but over time the money you save on your electricity bills pays you back for the initial investment.

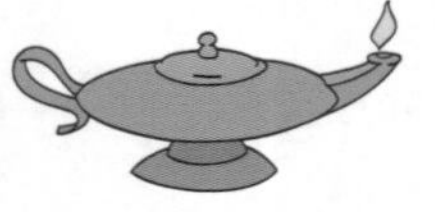
Not an energy efficient lamp.

Power Ratings of Appliances

The total energy transferred by an appliance depends on how long the appliance is on and its power rating.
The power of an appliance is the energy that it uses per second.

Energy Transferred = Power rating × time

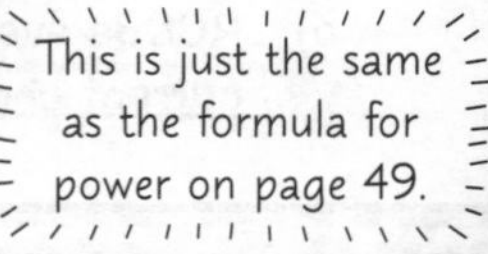

For example, if a 2.5 kW kettle is on for 5 minutes, the energy transferred by the kettle in this time is $300 \times 2500 = 750\,000$ J = 750 kJ. (5 minutes = 300 s).

This is just the same as the formula for power on page 49.

Ohm's girlfriend was a vixen — he couldn't resistor...

The equation for power is a real simple one, but it's absolutely essential that you've got it hard-wired into your memory. Remember: power is energy transferred per second. Power is energy transferred per second. Power is energy transferred per second....

Power and Energy Change

You can think about electrical circuits in terms of energy transfer — the charge carriers take charge around the circuit, and when they go through an electrical component energy is transferred to make the component work.

Electrical Power and Fuse Ratings

1) The formula for electrical power is: **POWER = CURRENT × POTENTIAL DIFFERENCE** $P = I \times V$

2) Most electrical goods show their power rating and voltage rating. To work out the size of the fuse needed, you need to work out the current that the item will normally use:

EXAMPLE: A hair dryer is rated at 230 V, 1 kW. Find the fuse needed.
ANSWER: I = P/V = 1000/230 = 4.3 A. Normally, the fuse should be rated just a little higher than the normal current, so a 5 amp fuse is ideal for this one.

The Potential Difference is the Energy Transferred per Charge Passed

1) When an electrical charge (Q) goes through a change in potential difference (V), then energy (E) is transferred.
2) Energy is supplied to the charge at the power source to 'raise' it through a potential.
3) The charge gives up this energy when it 'falls' through any potential drop in components elsewhere in the circuit.

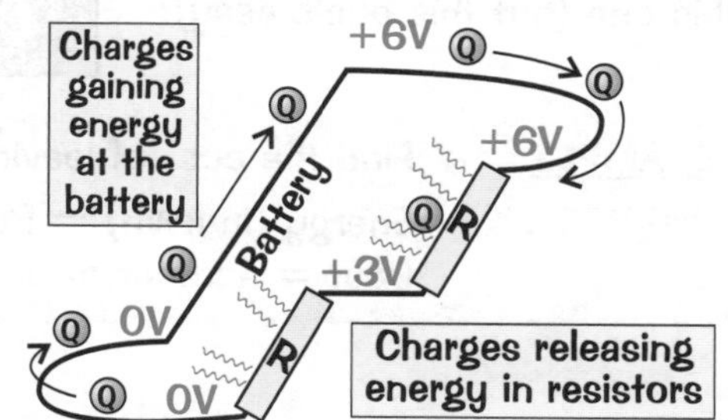

The formula is real simple:

Energy transformed = Charge × Potential difference

4) The bigger the change in P.D. (or voltage), the more energy is transferred for a given amount of charge passing through the circuit.
5) That means that a battery with a bigger voltage will supply more energy to the circuit for every coulomb of charge which flows round it, because the charge is raised up "higher" at the start (see above diagram) — and as the diagram shows, more energy will be dissipated in the circuit too.

EXAMPLE: The motor in an electric toothbrush is attached to a 3 V battery. If a current of 0.8 A flows through the motor for 3 minutes:

a) Calculate the total charge passed.
b) Calculate the energy transformed by the motor.
c) Explain why the kinetic energy output of the motor will be less than your answer to b).

ANSWER: a) Use the formula (p.57) Q = I × t = 0.8 × (3 × 60) = 144 C
b) Use E = Q × V = 144 × 3 = 432 J
c) The motor won't be 100% efficient. Some of the energy will be transformed into sound and heat.

You have the power — now use your potential...

Ok, another two formulas. By this point you're probably experiencing a little bit of formula fatigue, but trust me, you will be glad that you learned them all. Try to think about exactly what each one means and how they work together — things are a lot easier to memorise if you have a real understanding of why they are there.

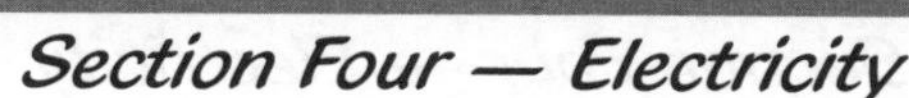

The Cost of Electricity

Isn't electricity great — generally, I mean. You can power all sorts of toys and gadgets with electricity. But it'll cost you. 'How much?' I hear you cry... Read and learn.

Kilowatt-hours (kWh) are "UNITS" of Energy

Your electricity meter records how much energy you use in units of kilowatt-hours, or kWh.

> A KILOWATT-HOUR is the amount of electrical energy converted by a 1 kW appliance left on for 1 HOUR.

The higher the power rating of an appliance, and the longer you leave it on, the more energy it consumes, and the more it costs. Learn (and practise rearranging) this equation too...

UNITS OF ENERGY (in kWh) = POWER (in kW) × TIME (in hours)

And this one (but this one's easy):

COST = NUMBER OF UNITS × PRICE PER UNIT

EXAMPLE: Find the cost of leaving a 60 W light bulb on for 30 minutes if one kWh costs 10p.

ANSWER: Energy (in kWh) = Power (in kW) × Time (in hours) = 0.06 kW × ½ hr = 0.03 kWh
Cost = number of units × price per unit = 0.03 × 10p = 0.3p

Off-Peak Electricity is Cheaper

Electricity supplied during the night (off-peak) is sometimes cheaper. Storage heaters (see p.15) take advantage of this — they heat up at night and then release the heat slowly throughout the day. If you can put washing machines, dishwashers, etc. on at night, so much the better.

ADVANTAGES of using off-peak electricity

1) Cost-effective for the electricity company — power stations can't be turned off at night, so it's good if there's a demand for electricity at night.
2) Cheaper for consumers if they buy electricity during the off-peak hours.

DISADVANTAGES of using off-peak electricity

1) There's a slightly increased risk of fire with more appliances going at night but no one watching.
2) You start fitting your routine around the cheap rate hours — i.e. you might stop enjoying the use of electricity during the day.

How to Read an Electricity Meter

1) Being able to read an electricity meter is an essential life skill — but don't worry, it's pretty straightforward. The units are usually in kWh — but make sure you check.
2) You could be given two meter readings and be asked to work out the total energy that's been used over a particular time period. Just subtract the meter reading at the start of the time (the smaller one) from the reading at the end to work this out.

Watt's the answer — well, part of it...

Get a bit of practice with the equations in those lovely bright red boxes, and try these questions:

1) A kettle uses 5.52 kWh of energy over a period of two hours. Calculate its power rating.
2) With 0.5 kWh of energy, for how long could you run the kettle?

Answers on p.140.

Choosing Electrical Appliances

Unfortunately, this isn't about what colour MP3 player to get...

Sometimes You Have a Choice of Electrical Equipment

There are often a few different appliances that do the same job. You should be able to weigh up the pros and cons of different appliances and decide which one is most suitable for a particular situation. Some appliances use less energy or are more cost-effective than others, and there are practical advantages and disadvantages to consider as well — e.g. 'Can an appliance be used in areas with limited electricity supplies?'

Here are some examples of the sorts of things you might have to compare:

E.G. CLOCKWORK RADIOS AND BATTERY RADIOS

1) Battery radios and clockwork radios are both handy in areas where there is no mains electricity supply.
2) Clockwork radios work by storing elastic potential energy in a spring when someone winds them up. The elastic potential energy is slowly released and used to power the radio.
3) Batteries can be expensive, but powering a clockwork radio is free.
4) Battery power is also only useful if you can get hold of some new batteries when the old ones run out. You don't get that problem with clockwork radios — but it can get annoying to have to keep winding them up every few hours to recharge them.
5) Clockwork radios are also better for the environment — a lot of energy and harmful chemicals go into making batteries, and they're often tricky to dispose of safely.

You Might Be Asked to Use Data to Compare Two Appliances

EXAMPLE

A company is deciding whether to install a 720 W low-power heater, or a high-power 9 kW heater. The heater they choose will be on for 30 hours each week. Their electricity provider charges 7p per kWh of electricity. How much money per week would they save by choosing the low-power heater?

ANSWER: Weekly electricity used by the low-power heater = 0.720 kW × 30 h = 21.6 kWh
Weekly electricity used by the high-power heater = 9 kW × 30 h = 270 kWh
Total saving = (270 – 21.6) × 7 = £17.39 (to the nearest penny)

Standard of Living is Affected by Access to Electricity

1) Most people in developed countries have access to mains electricity. However, many people living in the world's poorest countries don't — this has a big effect on their standard of living.
2) In the UK, our houses are full of devices that transform electrical energy into other useful types of energy. For example, not only is electric lighting useful and convenient, but it can also help improve safety at night.
3) Refrigerators keep food fresh for longer by slowing down the growth of bacteria. Refrigerators are also used to keep vaccines cold. Without refrigeration it's difficult to distribute important vaccines — this can have devastating effects on a country's population.
4) Electricity also plays an important role in improving public health in other ways. Hospitals in developed countries rely heavily on electricity, e.g. for X-ray machines. Without access to these modern machines, the diagnosis and treatment of patients would be poorer and could reduce life expectancy.
5) Communications are also affected by a lack of electricity. No electricity means no internet or phones — making it hard for people to keep in touch, or for people to send and receive news and information.

I'm definitely a fan of things running like clockwork...

Make sure you're happy with comparing electrical devices, and you know how important access to electricity can be.

Revision Summary for Section Four

It's business time — another chance for you to see which bits went in and which bits you need to flick back and have another read over. You know the drill by now. Do as many of the questions as you can and then try the tricky ones after you've had another chance to read the pages you struggled on. You know it makes sense.

1) What causes the build-up of static electricity? Which particles move when static builds up?
2) Describe the forces between objects with: a) like charges, b) opposite charges.
3) Explain how static electricity can make synthetic clothes crackle when you take them off.
4) Describe the dangers associated with static electricity when refuelling a vehicle. What can be done to make refuelling safer?
5) True or false: the greater the resistance of an electrical component, the smaller the current that flows through it?
6)* 240 C of charge is carried though a wire in a circuit in one minute. How much current has flowed through the wire?
7) What is another name for potential difference?
8) What formula relates work done, potential difference and charge?
9) Draw a diagram of the circuit that you would use to find the resistance of a motor.
10) Sketch typical potential difference-current graphs for: a) a resistor, b) a filament lamp, c) a diode. Explain the shape of each graph.
11) Explain how resistance of a component changes with its temperature in terms of ions and electrons.
12)* What potential difference is required to push 2 A of current through a 0.6 Ω resistor?
13)* Calculate the resistance of a wire if the potential across it is 12 V and the current through it is 2.5 A.
14) Give three applications of LEDs.
15) Describe how the resistance of an LDR varies with light intensity. Give an application of an LDR.
16)* A 4 Ω bulb and a 6 Ω bulb are connected in series with a 12 V battery.
    a) How much current flows through the 4 Ω bulb?
    b) What is the potential difference over the 6 Ω bulb?
    c) What would the potential difference over the 6 Ω bulb be if the two bulbs were connected in parallel?
17) Explain, in terms of energy, why P.D. is shared out in a series circuit.
18) Two circuits each contain a 2 Ω and a 4 Ω resistor — in one circuit they're in series, in the other they're in parallel. Which circuit will have the higher total resistance? Why?
19)* An AC supply of electricity has a time period of 0.08s. What is its frequency?
20) Name the three wires in a three-core cable.
21) Sketch and label a properly wired three-pin plug.
22) Explain fully how a fuse and earth wire work together.
23) How does an RCCB stop you from getting electrocuted?
24) What does the power of an appliance measure?
25)* Which uses more energy, a 45 W pair of hair straighteners used for 5 minutes, or a 105 W hair dryer used for 2 minutes?
26)* Calculate the energy transformed by a torch using a 6 V battery when 530 C of charge pass through.
27)* Calculate how many kWh of electrical energy are used by a 0.5 kW heater used for 15 minutes.
28) Would a battery-powered radio or a clockwork radio be more suitable to use in rural Africa? Why?

*Answers on page 140.

Atomic Structure

Ernest Rutherford didn't just pick the nuclear model of the atom out of thin air. It all started with a Greek fella called Democritus in the 5th Century BC. He thought that all matter, whatever it was, was made up of identical lumps called "atomos". And that's about as far as the theory got until the 1800s...

Rutherford Scattering and the Demise of the Plum Pudding

1) In 1804 John Dalton agreed with Democritus that matter was made up of tiny spheres ("atoms") that couldn't be broken up, but he reckoned that each element was made up of a different type of "atom".
2) Nearly 100 years later, J J Thomson discovered that electrons could be removed from atoms. So Dalton's theory wasn't quite right (atoms could be broken up). Thomson suggested that atoms were spheres of positive charge with tiny negative electrons stuck in them like plums in a plum pudding.
3) That "plum pudding" theory didn't last very long though. In 1909 Rutherford and Marsden tried firing a beam of alpha particles (see p.73) at thin gold foil. They expected that the positively charged alpha particles would be slightly deflected by the electrons in the plum pudding model.
4) However, most of the alpha particles just went straight through, but the odd one came straight back at them, which was frankly a bit of a shocker for Rutherford and his pal.

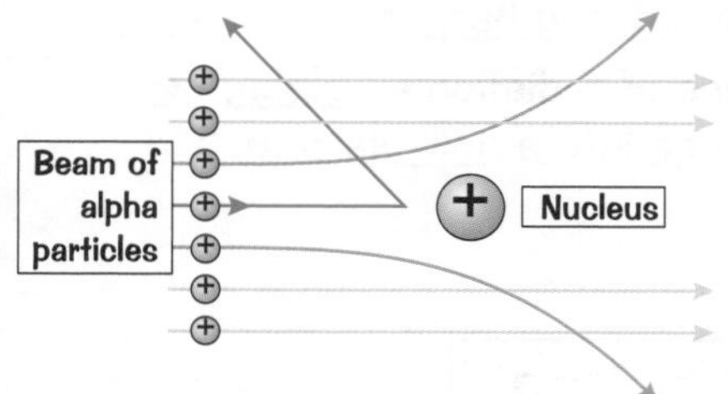

5) Being pretty clued-up guys, Rutherford and Marsden realised this meant that most of the mass of the atom was concentrated at the centre in a tiny nucleus. They also realised that the nucleus must have a positive charge, since it repelled the positive alpha particles.
6) It also showed that most of an atom is just empty space, which is also a bit of a shocker when you think about it.

Rutherford and Marsden Came Up with the Nuclear Model of the Atom

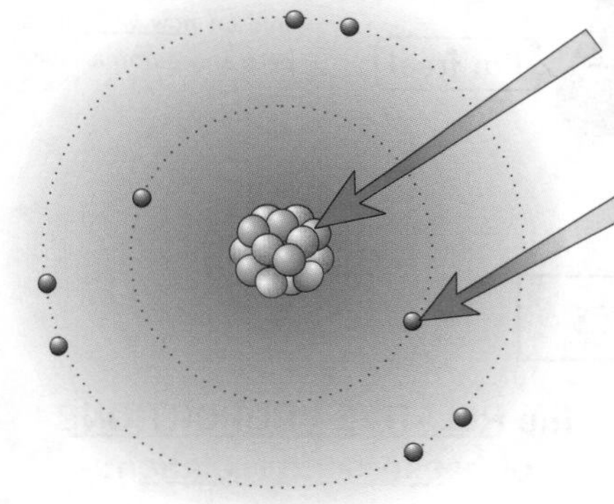

The nucleus is tiny but it makes up most of the mass of the atom. It contains protons (which are positively charged) and neutrons (which are neutral) — which gives it an overall positive charge.

The rest of the atom is mostly empty space. The negative electrons whizz round the outside of the nucleus really fast. They give the atom its overall size — the radius of the atom's nucleus is about 10 000 times smaller than the radius of the atom. Crikey.

Here are the relative charges and masses of each particle:

PARTICLE	MASS	CHARGE
Proton	1	+1
Neutron	1	0
Electron	$1/2000$	- 1

Number of Protons Equals Number of Electrons

1) Atoms have no charge overall.
2) The charge on an electron is the same size as the charge on a proton — but opposite.
3) This means the number of protons always equals the number of electrons in a neutral atom.
4) If some electrons are added or removed, the atom becomes a charged particle called an ion.

And I always thought Kate Moss was the best model...

The nuclear model is just one way of thinking about the atom. It works really well for explaining a lot of physical properties of different elements, but it's certainly not the whole story. Other bits of science are explained using different models of the atom. The beauty of it though is that no one model is more right than the others.

Atoms and Ionising Radiation

You have just entered the subatomic realm — now stuff starts to get real interesting...

Isotopes are Different Forms of the Same Element

1) Isotopes are atoms with the same number of protons but a different number of neutrons.
2) Hence they have the same atomic number, but different mass numbers.
3) Atomic (proton) number is the number of protons in an atom.
4) Mass (nucleon) number is the number of protons + the number of neutrons in an atom.
5) Carbon-12 and carbon-14 are good examples of isotopes:
6) Most elements have different isotopes, but there's usually only one or two stable ones.
7) The other isotopes tend to be unstable.

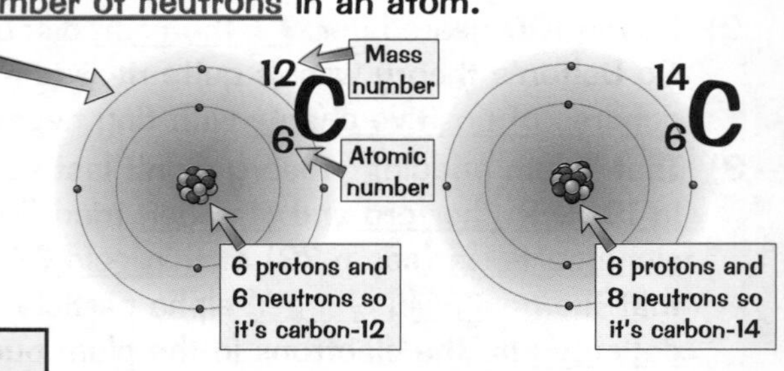

Radioactivity is a Totally Random Process

1) Unstable nuclei are radioactive — they will decay and in the process give out ionising radiation (see below).
2) This process is entirely random. This means that if you have 1000 unstable nuclei, you can't say when any one of them is going to decay, nor can you do anything at all to make a decay happen.
3) When the nucleus does decay it will spit out one or more of three types of radiation — alpha, beta or gamma (see next page). In the process, the nucleus will often change into a new element.

Alpha, Beta and Gamma Radiation can Cause Ionisation

1) Atoms can gain or lose electrons. When an atom (with no overall charge) loses or gains an electron it is turned into an ion (which is charged). This is known as ionisation (see p.35).
2) Alpha, beta and gamma are all types of ionising radiation — they can cause ionisation of atoms.

See the next page for info on how well alpha, beta and gamma ionise.

Background Radiation Comes from Many Sources

The background radiation we receive comes from:

1) Radioactivity of naturally occurring unstable isotopes which are all around us — in the air, in food, in building materials and in the rocks under our feet. A large proportion of background radiation comes from these natural sources.
2) Radiation from space, which is known as cosmic rays. These come mostly from the Sun.
3) Radiation due to human activity, e.g. fallout from nuclear explosions, or waste from industry and hospitals. But this represents a small proportion of the total.
4) The amount of background radiation can vary depending on where you are and your job (see page 74).

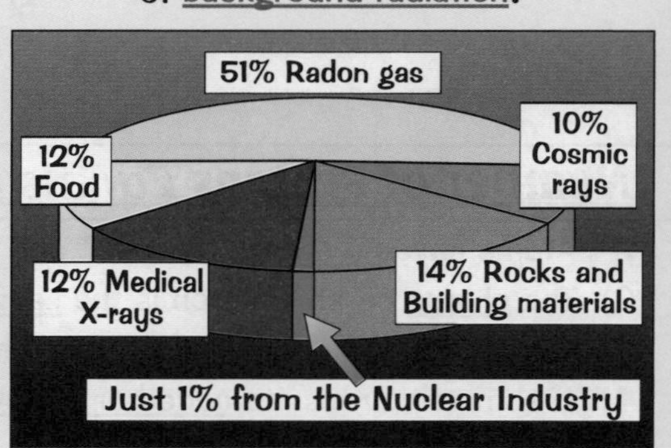

Completely random — just like your revision shouldn't be...

It's the number of protons which decides what element something is, then the number of neutrons decides what isotope of that element it is. And it's unstable isotopes which undergo radioactive decay.

Ionising Radiation

Alpha (α) Beta (β) Gamma (γ) — there's a short alphabet of radiation for you to learn here. And it's all ionising.

Alpha Particles are Helium Nuclei

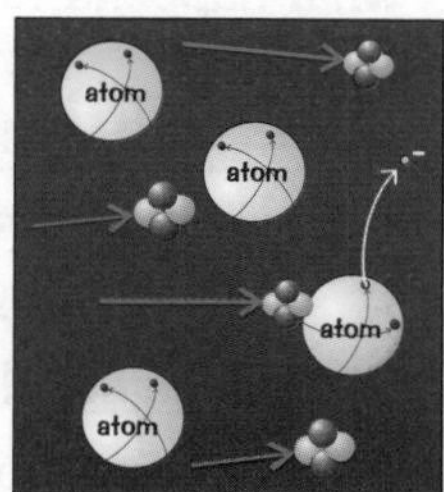

1) An alpha particle is two neutrons and two protons — the same as a helium nucleus.
2) Alpha particles have a mass of 4 and a charge of +2.
3) They are relatively big and heavy and slow moving.
4) They therefore don't penetrate very far into materials and are stopped quickly, even when travelling through air.
5) Because of their size they are strongly ionising, which just means they bash into a lot of atoms and knock electrons off them before they slow down, which creates lots of ions.

Beta Particles are Electrons

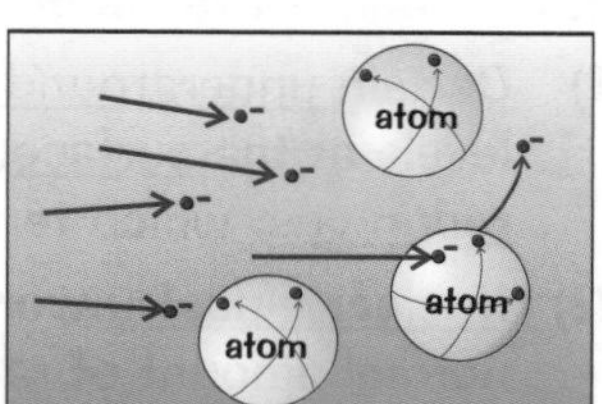

1) Beta particles are in between alpha and gamma in terms of their properties.
2) They move quite fast and they are quite small (they're electrons).
3) They penetrate moderately into materials before colliding, have a long range in air, and are moderately ionising too.
4) For every β-particle emitted, a neutron turns to a proton in the nucleus.
5) A β-particle is simply an electron, with virtually no mass and a charge of –1.

Nuclear Equations Need to Balance

1) You can write alpha and beta decays as nuclear equations.
2) Watch out for the mass and atomic numbers — they have to balance up on both sides.

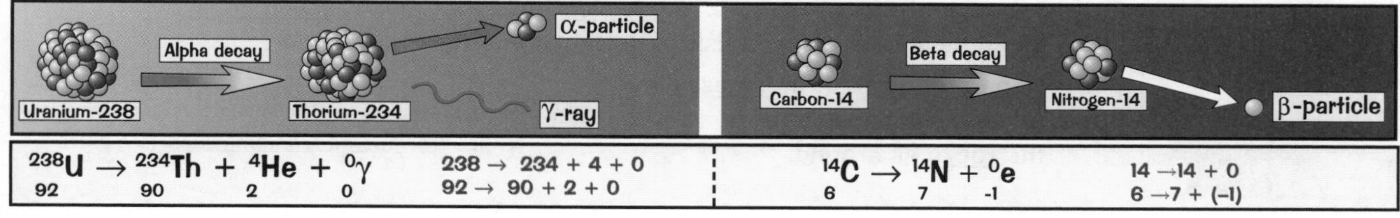

$^{238}_{92}U \rightarrow {}^{234}_{90}Th + {}^{4}_{2}He + {}^{0}_{0}\gamma$ 238 → 234 + 4 + 0, 92 → 90 + 2 + 0

$^{14}_{6}C \rightarrow {}^{14}_{7}N + {}^{0}_{-1}e$ 14 → 14 + 0, 6 → 7 + (–1)

Gamma Rays are Very Short Wavelength EM Waves

1) After spitting out an alpha or beta particle, the nucleus might need to get rid of some extra energy.
2) It does this by emitting a gamma ray — a type of electromagnetic wave.
3) Gamma rays penetrate far into materials without being stopped and pass straight through air.
4) This means they are weakly ionising because they tend to pass through rather than collide with atoms. Eventually they hit something and do damage.
5) Gamma rays have no mass and no charge.
6) Since a gamma ray is just energy, it doesn't change the element of the nucleus that emits it.

Remember What Blocks the Three Types of Radiation...

Alpha particles are blocked by paper.

Beta particles are blocked by thin aluminium.

Gamma rays are blocked by thick lead.

Similar things will also block them, e.g. skin will stop alpha, a thin sheet of any metal will stop beta, and very thick concrete will stop gamma.

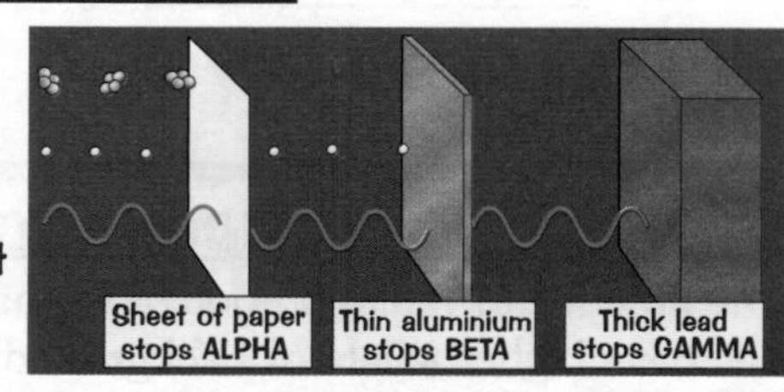

Ionising Radiation

Ooh, it's a mixed bag this page. First up, what affects how much radiation we're exposed to. Then a look at what happens to alpha and beta particles in magnetic and electric fields... And a partridge in a pear treeeeee.

The Damage Caused By Radiation Depends on the Radiation Dose

How likely you are to suffer damage if you're exposed to nuclear radiation depends on the radiation dose.

1) Radiation dose depends on the type and amount of radiation you've been exposed to.
2) The higher the radiation dose, the more at risk you are of developing cancer (see p.77)
3) Radiation dose is measured in sieverts (Sv) or more usually millisieverts (mSv).

Radiation Dose Depends on Location and Occupation

The amount of radiation you're exposed to can be affected by your location and occupation:

Coloured bits indicate more radiation from rocks

1) Certain underground rocks (e.g. granite) can cause higher levels at the surface, especially if they release radioactive radon gas, which tends to get trapped inside people's houses.
2) Nuclear industry workers and uranium miners are typically exposed to 10 times the normal amount of radiation. They wear protective clothing and face masks to stop them from touching or inhaling the radioactive material, and monitor their radiation doses with special radiation badges and regular check-ups.
3) Radiographers work in hospitals using ionising radiation and so have a higher risk of radiation exposure. They wear lead aprons and stand behind lead screens to protect them from prolonged exposure to radiation.

4) At high altitudes (e.g. in jet planes) the background radiation increases because of more exposure to cosmic rays. That means commercial pilots have an increased risk of getting some types of cancer.
5) Underground (e.g. in mines, etc.) it increases because of the rocks all around.

Radioactive materials put people at risk through either irradiation or contamination (see p.19).

Alpha and Beta Particles are Deflected by Electric and Magnetic Fields

1) Alpha particles have a positive charge, beta particles have a negative charge.
2) When travelling through a magnetic or electric field, both alpha and beta particles will be deflected.
3) They're deflected in opposite directions because of their opposite charge.
4) Alpha particles have a larger charge than beta particles, and feel a greater force in magnetic and electric fields. But they're deflected less because they have a much greater mass.
5) Gamma radiation is an electromagnetic (EM) wave and has no charge, so it doesn't get deflected by electric or magnetic fields.

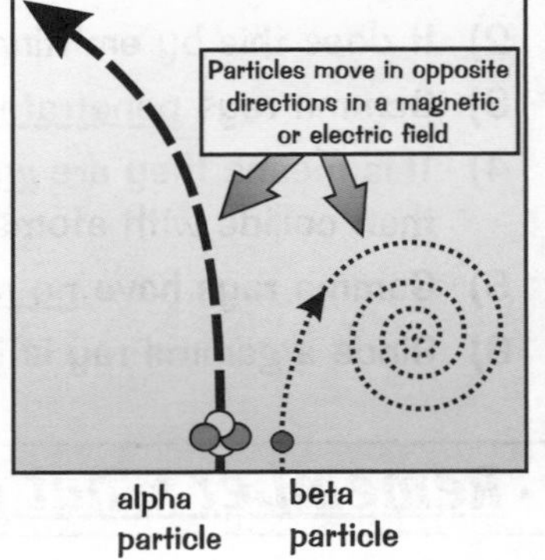

Beta particles in magnetic field? This page has spiralled out of control...

So the amount of radiation you're exposed to depends on your job and your location. Don't forget that some places have higher levels of background radiation than others — so the people there'll get a higher radiation dose.

Half-Life

The unit for measuring radioactivity is the becquerel (Bq). 1 Bq means one nucleus decaying per second.

The Radioactivity of a Sample Always Decreases Over Time

1) This is pretty obvious when you think about it. Each time a decay happens and an alpha, beta or gamma is given out, it means one more radioactive nucleus has disappeared.
2) Obviously, as the unstable nuclei all steadily disappear, the activity (the number of nuclei that decay per second) will decrease. So the older a sample becomes, the less radiation it will emit.

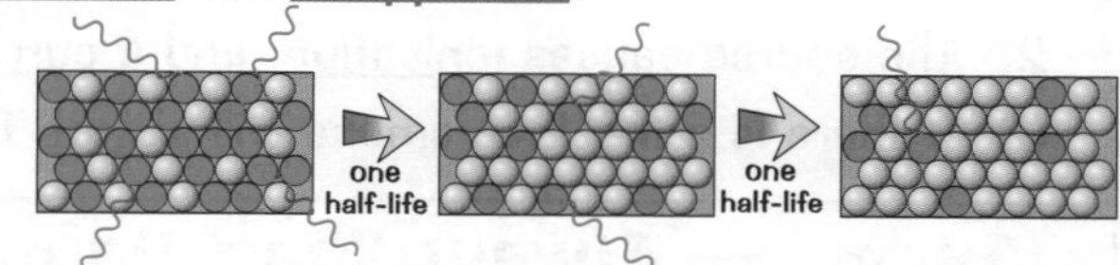

3) How quickly the activity drops off varies a lot. For some substances it takes just a few microseconds before nearly all the unstable nuclei have decayed, whilst for others it can take millions of years.
4) The problem with trying to measure this is that the activity never reaches zero, which is why we have to use the idea of half-life to measure how quickly the activity drops off.
5) Learn this definition of half-life:

HALF-LIFE is the AVERAGE TIME it takes for the NUMBER OF NUCLEI in a RADIOACTIVE ISOTOPE SAMPLE to HALVE.

6) In other words, it is the time it takes for the count rate (the number of radioactive emissions detected per unit of time) from a sample containing the isotope to fall to half its initial level.
7) A short half-life means the activity falls quickly, because lots of the nuclei decay quickly.
8) A long half-life means the activity falls more slowly because most of the nuclei don't decay for a long time — they just sit there, basically unstable, but kind of biding their time.

Do Half-life Questions Step by Step

Half-life is maybe a little confusing, but exam calculations are straightforward so long as you do them slowly, STEP BY STEP. Like this one:

A VERY SIMPLE EXAMPLE: The activity of a radioisotope is 640 cpm (counts per minute). Two hours later it has fallen to 80 cpm. Find the half-life of the sample.

ANSWER: You must go through it in short simple steps like this:

INITIAL count:		after ONE half-life:		after TWO half-lives:		after THREE half-lives:
640	(÷2)→	320	(÷2)→	160	(÷2)→	80

Notice the careful step-by-step method, which tells us it takes three half-lives for the activity to fall from 640 to 80. Hence two hours represents three half-lives, so the half-life is 120 mins ÷ 3 = 40 minutes.

Finding the Half-life of a Sample Using a Graph

1) The data for the graph will usually be several readings of count rate taken with a G-M tube and counter.
2) The graph will always be shaped like the one shown.
3) The half-life is found from the graph by finding the time interval on the bottom axis corresponding to a halving of the activity on the vertical axis. Easy peasy really.

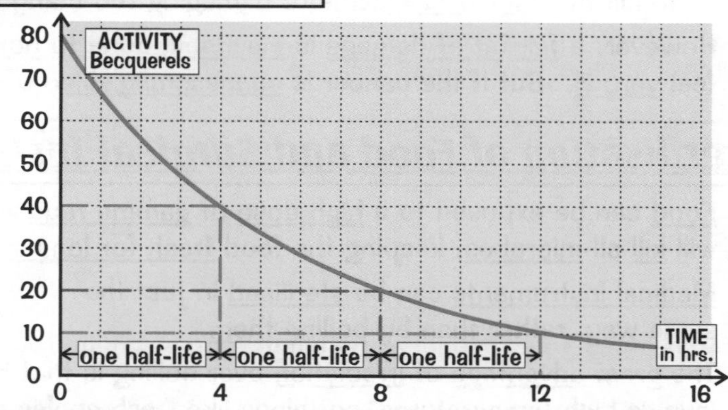

Half-life of a box of chocolates — about five minutes...

For medical applications, you need to use isotopes that have a suitable half-life. A radioactive tracer needs to have a short half-life to minimise the risk of damage to the patient. A source for sterilising equipment needs to have a long half-life, so you don't have to replace it too often (see next page).

Uses of Radiation

Radiation gets a lot of bad press, but the fact is it's essential for things like modern medicine. Read on chaps...

Smoke Detectors — Use α-Radiation

1) A weak source of alpha radiation is placed in the detector, close to two electrodes.
2) The source causes ionisation, and a current flows between the electrodes.
3) If there is a fire then smoke will absorb the radiation — so the current stops and the alarm sounds.

Tracers — Always Short Half-Life β or γ-Emitters

1) Certain radioactive isotopes can be used as medical tracers. They're injected into people (or they can just swallow them) and their progress around the body can be followed using an external detector. A computer converts the reading to a display showing where the strongest reading is coming from.
2) A well-known example is the use of iodine-131, which is absorbed by the thyroid gland just like normal iodine-127, but it gives out radiation which can be detected to indicate whether the thyroid gland is taking in iodine as it should.

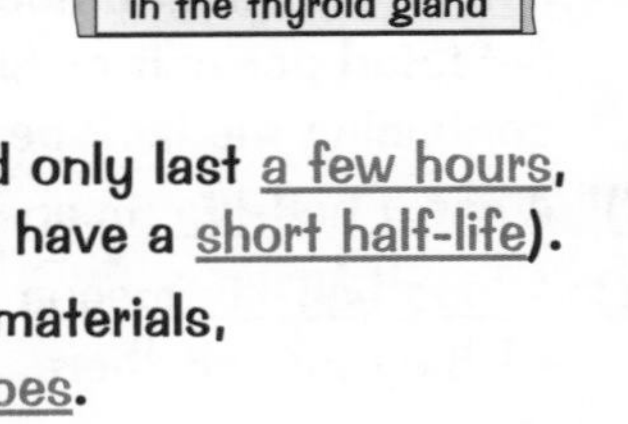

3) All isotopes which are taken into the body must be GAMMA or BETA emitters (never alpha), so that the radiation passes out of the body — and they should only last a few hours, so that the radioactivity inside the patient quickly disappears (i.e. they should have a short half-life).

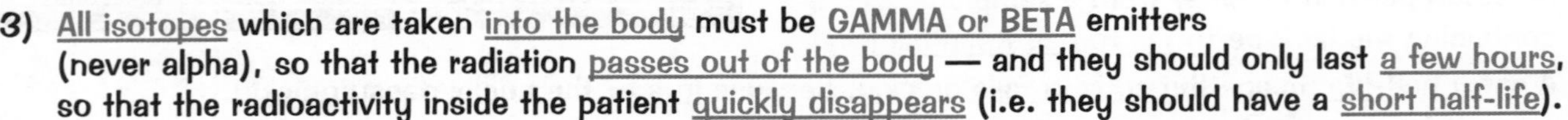

4) Gamma tracers can also be used in industry to track the movement of waste materials, find the route of underground pipe systems or detect leaks or blockages in pipes.

β-Radiation is Used in Thickness Gauges

1) Beta radiation is used in thickness control. You direct radiation through the stuff being made (e.g. paper), and put a detector on the other side, connected to a control unit. When the amount of detected radiation changes, it means the paper is coming out too thick or too thin, so the control unit adjusts the rollers to give the correct thickness.
2) The radioactive source used needs to have a fairly long half-life so it doesn't decay away too quickly.

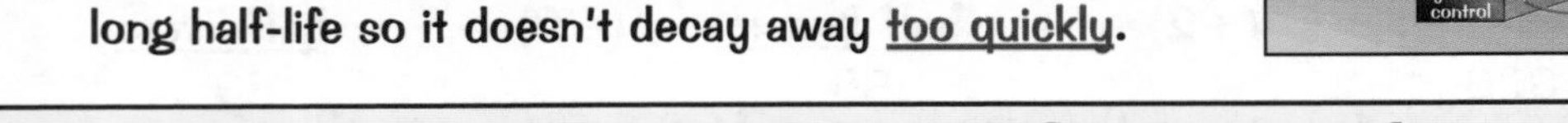

Radiotherapy — the Treatment of Cancer Using γ-Rays

1) Since high doses of gamma rays will kill all living cells, they can be used to treat cancers.
2) The gamma rays have to be directed carefully and at just the right dosage so as to kill the cancer cells without damaging too many normal cells.
3) However, a fair bit of damage is inevitably done to normal cells, which makes the patient feel very ill. But if the cancer is successfully killed off in the end, then it's worth it.

Sterilisation of Food and Surgical Instruments Using γ-Rays

1) Food can be exposed to a high dose of gamma rays which will kill all microbes, keeping the food fresh for longer.
2) Medical instruments can be sterilised in just the same way, rather than by boiling them.

3) The great advantage of irradiation over boiling is that it doesn't involve high temperatures, so things like fresh apples or plastic instruments can be totally sterilised without damaging them.
4) The food is not radioactive afterwards, so it's perfectly safe to eat.
5) The isotope used for this needs to be a very strong emitter of gamma rays with a reasonably long half-life (at least several months) so that it doesn't need replacing too often.

Radioactivity Safety

When Marie Curie discovered the radioactive properties of radium in 1898, nobody knew about its dangers. Radium was used to make glow-in-the-dark watches and many watch dial painters developed cancer as a result.

Radiation Harms Living Cells

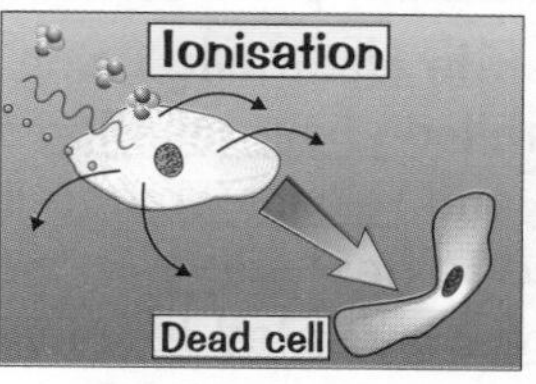

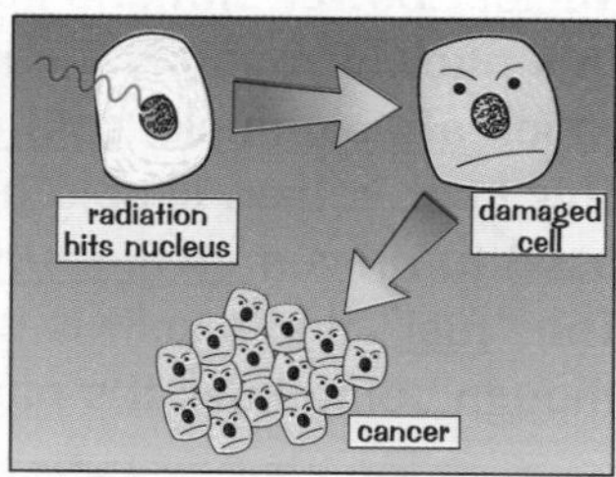

1) Alpha, beta and gamma radiation will cheerfully enter living cells and collide with molecules.
2) These collisions cause ionisation, which damages or destroys the molecules.
3) Lower doses tend to cause minor damage without killing the cell.
4) This can give rise to mutant cells which divide uncontrollably. This is cancer.
5) Higher doses tend to kill cells completely, which causes radiation sickness if a lot of body cells all get blatted at once.
6) The extent of the harmful effects depends on two things:
   a) How much exposure you have to the radiation.
   b) The energy and penetration of the radiation, since some types are more hazardous than others, of course.

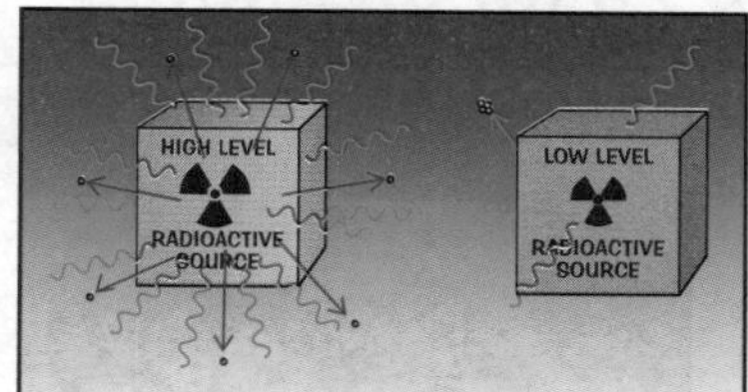

Outside the Body, β and γ–Sources are the Most Dangerous

This is because beta and gamma can get inside to the delicate organs, whereas alpha is much less dangerous because it can't penetrate the skin.

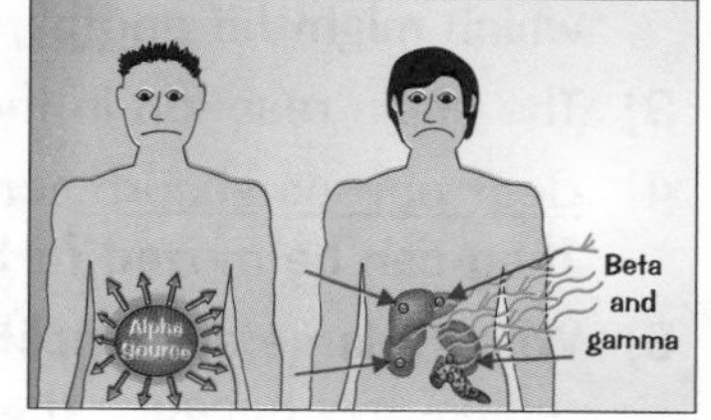

Inside the Body, an α-Source is the Most Dangerous

Inside the body alpha sources do all their damage in a very localised area. Beta and gamma sources on the other hand are less dangerous inside the body because they mostly pass straight out without doing much damage.

You Need to Learn About These Safety Precautions

Obviously radioactive materials need to be handled carefully. But in the exam they might ask you to evaluate some specific precautions that should be taken when handling radioactive materials.

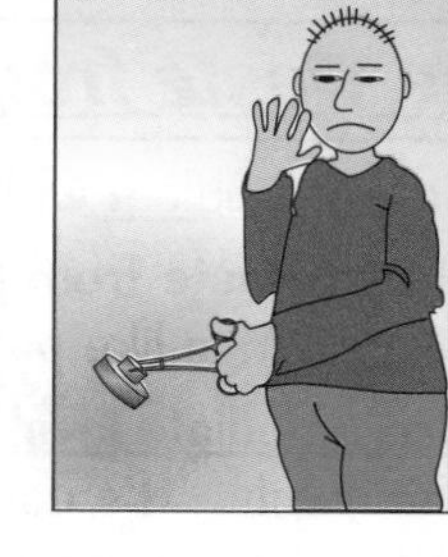

1) When conducting experiments, use radioactive sources for as short a time as possible so your exposure is kept to a minimum.
2) Never allow skin contact with a source. Always handle with tongs.
3) Hold the source at arm's length to keep it as far from the body as possible. This will decrease the amount of radiation that hits you, especially for alpha particles as they don't travel far in air.
4) Keep the source pointing away from the body and avoid looking directly at it.
5) Lead absorbs all three types of radiation (though a lot of it is needed to stop gamma radiation completely). Always store radioactive sources in a lead box and put them away as soon as the experiment is over. Medical professionals who work with radiation every day (such as radiographers) wear lead aprons and stand behind lead screens for extra protection because of its radiation absorbing properties.
6) When someone needs an X-ray or radiotherapy, only the area of the body that needs to be treated is exposed to radiation. The rest of the body is protected with lead or other radiation absorbing materials.

Radiation sickness — well yes, it does all get a bit tedious...

Sadly, much of our knowledge of the harmful effects of radiation has come as a result of devastating events such as the atomic bombing of Japan in 1945. In the months following the bombs, thousands suffered from radiation sickness — the symptoms of which include nausea, fatigue, skin burns, hair loss and, in serious cases, death. In the long term, the area has experienced increased rates of cancer, particularly leukaemia.

Nuclear Fission

With the right set-up you can generate some serious energy using unstable isotopes.

Nuclear Fission — the Splitting Up of Big Atomic Nuclei

Nuclear power stations generate electricity using nuclear reactors. In a nuclear reactor, a controlled chain reaction takes place in which atomic nuclei split up and release energy in the form of heat. This heat is then simply used to heat water to make steam, which is used to drive a steam turbine connected to an electricity generator. The "fuel" that's split is usually uranium-235, though sometimes it's plutonium-239 (or both).

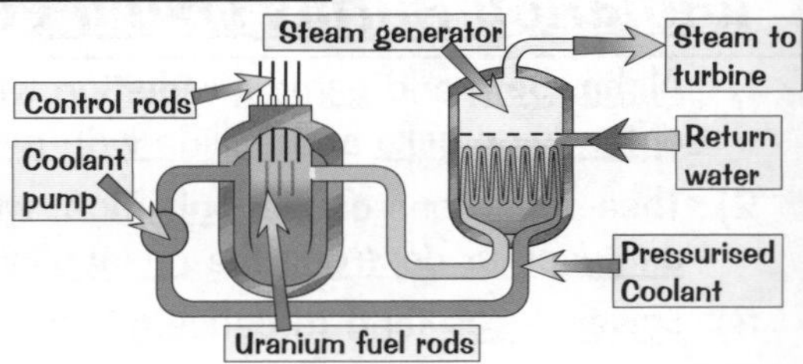

The Chain Reactions:

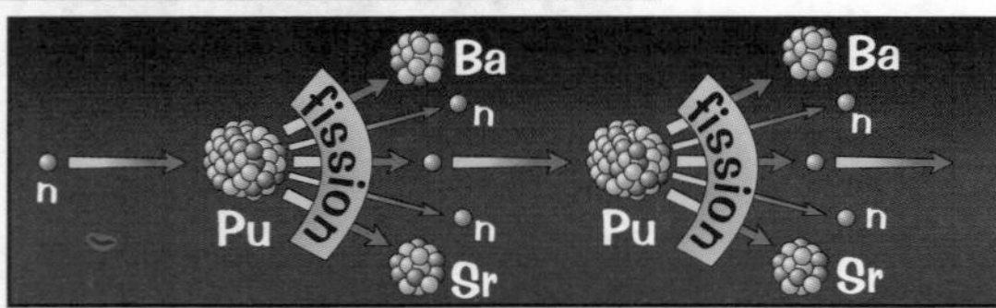

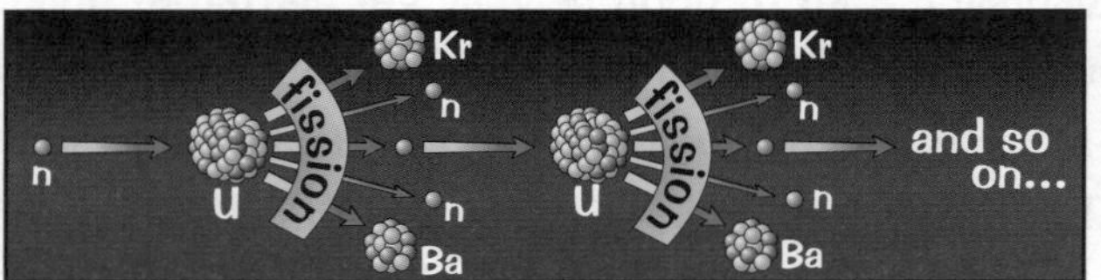

1) For nuclear fission to happen, a slow moving neutron must be absorbed into a uranium or plutonium nucleus. This addition of a neutron makes the nucleus unstable, causing it to split.
2) Each time a uranium or plutonium nucleus splits up, it spits out two or three neutrons, one of which might hit another nucleus, causing it to split also, and thus keeping the chain reaction going.
3) The chain reaction in the reactor has to be controlled, or the reactor would overheat.
4) Control rods absorb some of the neutrons and slow down the reaction. They can be moved further into and out of the reactor to absorb more or less neutrons.
5) When a large atom splits in two it will form two new smaller nuclei. These new nuclei are usually radioactive because they have the "wrong" number of neutrons in them.
6) A nucleus splitting (called a fission) gives out a lot of energy — lots more energy than you get from any chemical reaction. Nuclear processes release much more energy than chemical processes do. That's why nuclear bombs are so much more powerful than ordinary bombs (which rely on chemical reactions).

The Waste from Nuclear Power Stations is Hard to Deal With

The main problem with nuclear power is that it produces radioactive waste.

1) Most waste from power stations (or medical use, see p.76) is 'low level' (slightly radioactive). E.g. things like paper and gloves, etc. This waste can be disposed of by burying it in secure landfill sites.
2) Intermediate level waste includes things like the metal cases of used fuel rods and some waste from hospitals. It's usually quite radioactive — and some of it will stay that way for tens of thousands of years. It's often sealed into concrete blocks then put in steel canisters for storage.
3) High level waste from nuclear power stations is so radioactive that it generates a lot of heat. This waste is sealed in glass and steel, then cooled for about 50 years before it's moved to more permanent storage.
4) The canisters of intermediate and high level wastes could then be buried deep underground. However, it's difficult to find suitable places. The site has to be geologically stable (e.g. not suffer earthquakes), since big movements in the rock could break the canisters and radioactive material could leak out.
5) So, at the moment, most intermediate and high level waste is kept 'on-site' at nuclear power stations.
6) Nuclear fuel is cheap but the overall cost of nuclear power is high due to the cost of the power plant, waste processing and final decommissioning. Dismantling a nuclear plant safely takes decades.
7) Nuclear power also carries the risk of radiation leaks from the plant or a major catastrophe like Chernobyl.

I'm split over fissi drinks — loads of energy, but bad for your teeth...

There's enough uranium and plutonium around to provide us with energy for years, but it's difficult to deal with the waste we've already got. Learn about the three 'levels' of waste and how they're disposed of.

Nuclear Fusion

Loads of energy's released either when you break apart really big nuclei or join together really small nuclei. You can't do much with the ones in the middle, I'm afraid. (Don't ask, you don't want to know.)

Nuclear Fusion — The Joining of Small Atomic Nuclei

1) Nuclear fusion is the opposite of nuclear fission.
2) In nuclear fusion, two light nuclei combine to create a larger nucleus.
3) The most common example is two atoms of different hydrogen isotopes combining to form helium:

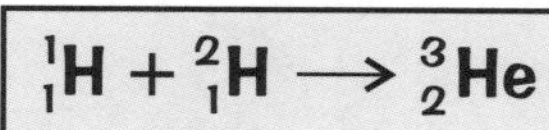

$${}^{1}_{1}H + {}^{2}_{1}H \rightarrow {}^{3}_{2}He$$

4) Fusion releases a lot of energy (more than fission for a given mass) — all the energy released in stars comes from fusion at extremely high temperatures and pressures. So people are trying to develop fusion reactors to make electricity.
5) Fusion doesn't leave behind much radioactive waste and there's plenty of hydrogen about to use as fuel.
6) The big problem is that fusion only happens at really high pressures and temperatures (about 10 000 000 °C).
7) No material can physically withstand that kind of temperature and pressure — so fusion reactors are really hard to build.
8) It's also hard to safely control the high temperatures and pressures.
9) There are a few experimental reactors around at the moment, the biggest one being JET (Joint European Torus), but none of them are generating electricity yet. It takes more power to get up to temperature than the reactor can produce.
10) Research into fusion power production is carried out by international groups to share the costs, expertise, experience and the benefits (when they eventually get it to work reliably).

FUSION BOMBS

- Fusion reactions also happen in fusion bombs.
- You might have heard of them as hydrogen, or H bombs.
- In fusion bombs, a fission reaction is used first to create the really high temperatures needed for fusion.

Cold Fusion — Hoax or Energy of the Future?

1) A new scientific theory has to go through a validation process before it's accepted.
2) An example of a theory which hasn't been accepted yet is 'cold fusion'.
3) Cold fusion is nuclear fusion which occurs at around room temperature, rather than at millions of degrees Celsius.
4) In 1989, two scientists reported that they had succeeded in releasing energy from cold fusion, using a simple experiment. This caused a lot of excitement — cold fusion would make it possible to generate lots of electricity, easily and cheaply.
5) After the press conference, the experiments and data were shared with other scientists so they could repeat the experiments. But few managed to reproduce the results reliably — so it hasn't been accepted as a realistic method of energy production.

Pity they can't release energy by confusion...*

Fusion bombs are incredibly powerful — they can release a few thousand times more energy than the nuclear fission bombs that destroyed Hiroshima and Nagasaki in World War II. Fusion power would be more useful...

*There'd be plenty of physics books to use as fuel.

Revision Summary for Section Five

There's some pretty heavy physics in this section. But just take it one page at a time and it's not so bad. You're even allowed to go back through the pages for a sneaky peak if you get stuck on these questions.

1) Explain how the experiments of Rutherford and Marsden led to the nuclear model of the atom.
2) Draw a table stating the relative mass and charge of the three basic subatomic particles.
3) Explain what isotopes are. Give an example. Do stable or unstable isotopes undergo nuclear decay?
4) True or false: radioactive decay can be triggered by certain chemical reactions.
5) What is ionisation?
6) Give three sources of background radiation.
7) What type of subatomic particle is a beta particle?
8)* Complete the following nuclear equations by working out the missing numbers shown by the dotted lines:
a) $^{131}_{53}I \rightarrow {}^{\dots}_{\dots}Xe + {}^{0}_{-1}\beta$ b) $^{\dots}_{\dots}Gd \rightarrow {}^{144}_{62}Sm + {}^{4}_{2}\alpha$
9) What substances could be used to block:
a) alpha radiation, b) beta radiation, c) gamma radiation?
10) What are the units of radiation dose? What two things does radiation 'dose' take into account?
11) List two places where the level of background radiation is increased and explain why.
12) Name three occupations that have an increased risk of exposure to radiation.
13) Sketch the paths of an alpha particle and a beta particle travelling through an electric field.
14) Define half-life.
15)* The activity of a radioactive sample is 840 Bq. Four hours later it has fallen to 105 Bq.
Find the half-life of the sample.
16) Sketch a typical graph of activity against time for a radioactive source.
Show how you can find the half-life from your graph.
17) Describe in detail how radioactive sources are used in each of the following:
a) medical tracers, b) treating cancer, c) sterilising food, d) sterilising medical equipment.
18) Explain what kind of damage ionising radiation causes to body cells. What are the effects of high doses? What damage can lower doses do?
19) Which is the most dangerous form of radiation if you eat it? Why?
20) Describe the precautions you should take when handling radioactive sources in the laboratory.
21) Draw a diagram to illustrate the fission of uranium-235 and explain how the chain reaction works.
22) What is the main environmental problem associated with nuclear power?
23) What is nuclear fusion? Why is it difficult to construct a working fusion reactor?
24) Briefly explain why cold fusion isn't accepted as a realistic method of energy production.

*Answers on page 140.

Magnetic Fields

There's a proper definition of a magnetic field:

> A MAGNETIC FIELD is a region where MAGNETIC MATERIALS (like iron and steel) and also WIRES CARRYING CURRENTS experience A FORCE acting on them.

Magnetic fields can be represented by field diagrams (e.g. see coil of wire diagram below).
The arrows on the field lines always point FROM THE NORTH POLE of the magnet TO THE SOUTH POLE!

The Magnetic Field Round a Current-Carrying Wire

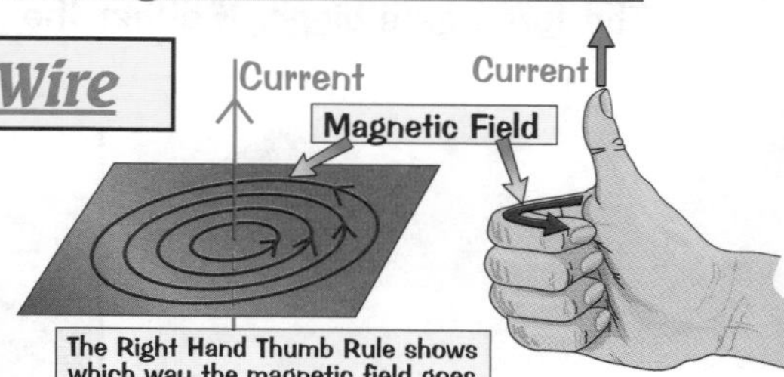

The Right Hand Thumb Rule shows which way the magnetic field goes

1) When a current flows through a wire, a magnetic field is created around the wire.
2) The field is made up of concentric circles with the wire in the centre.

A Rectangular Coil Reinforces the Magnetic Field

1) If you bend the current-carrying wire round into a coil, the magnetic field looks like this.
2) The circular magnetic fields around the sides of the loop reinforce each other at the centre.
3) If the coil has lots of turns, the magnetic fields from all the individual loops reinforce each other even more.

Current
Magnetic field

The Magnetic Field Round a Coil of Wire

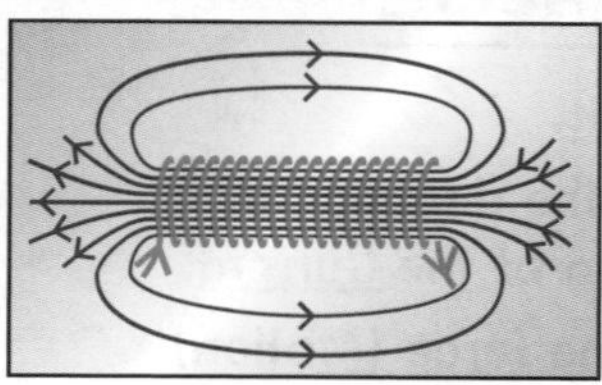

1) The magnetic field inside a coil of wire (a solenoid) is strong and uniform.
2) Outside the coil, the magnetic field is just like the one round a bar magnet.
3) You can increase the strength of the magnetic field around a solenoid by adding a magnetically "soft" iron core through the middle of the coil. It's then called an ELECTROMAGNET.

A magnetically soft material magnetises and demagnetises very easily. So, as soon as you turn off the current through the solenoid, the magnetic field disappears — the iron doesn't stay magnetised. This is what makes it useful for something that needs to be able to switch its magnetism on and off (see below).

Electromagnets are Useful as Their Magnetism can be Switched Off

An electromagnet must be constantly supplied with current — as that's what produces the magnetic field. So if the current stops, then it stops being magnetic. Magnets you can switch off at your whim can be really useful...

EXAMPLE: Cranes used for lifting iron and steel

1) Magnets can be used to attract and pick up things made from magnetic materials like iron and steel.
2) Electromagnets are used in some cranes, e.g. in scrap yards and steel works.
3) If an ordinary magnet was used, the crane would be able to pick up the cars etc., but then wouldn't let it go. Which isn't very helpful.
4) Using an electromagnetic means the magnet can be switched on when you want to and attract and pick stuff up, then switched off when you want to drop it. Which is far more useful.

I'm magnetically soft — I always cry when electromagnets are turned off...

Electromagnets pop up in lots of different places, e.g. electric bells, car ignition circuits, security doors. How strong they are depends on stuff like the number of turns on the coil and the size of current going through it.

The Motor Effect

Passing an electric current through a wire produces a magnetic field around the wire (p.81). If you put that wire into a magnetic field, you have two magnetic fields combining, which puts a force on the wire (generally).

A Current in a Magnetic Field Experiences a Force

The force experienced by a current-carrying wire in a magnetic field is known as the motor effect.

1) The two tests below demonstrate the force on a current-carrying wire placed in a magnetic field. The force gets bigger if either the current or the magnetic field is made bigger.

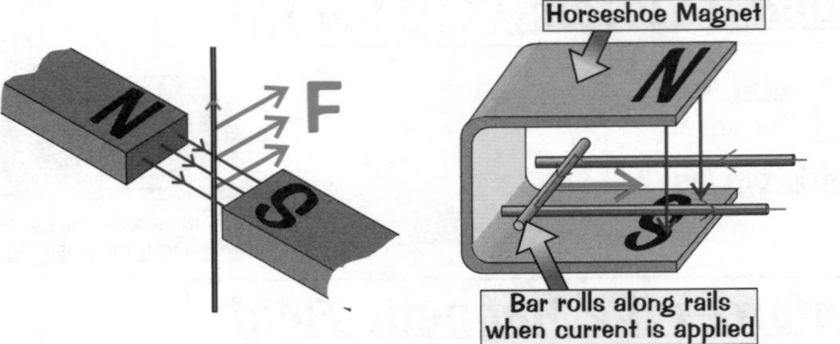

The motor effect is used in lots of appliances that use movement — see the next page.

2) Note that in both cases the force on the wire is at 90° to both the wire and to the magnetic field.
3) If the direction of the current or magnetic field is reversed, then the direction of the force is reversed too. You can always predict which way the force will act using Fleming's left hand rule as shown below.
4) To experience the full force, the wire has to be at 90° to the magnetic field.
5) If the wire runs along parallel to the magnetic field it won't experience any force at all.
6) At angles in between it'll feel some force.

Fleming's Left Hand Rule Tells You Which Way the Force Acts

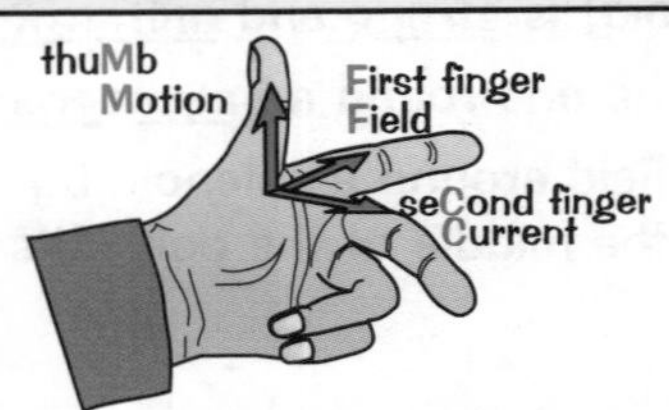

1) They could test if you can do this, so practise it.
2) Using your left hand, point your First finger in the direction of the Field and your seCond finger in the direction of the Current.
3) Your thuMb will then point in the direction of the force (Motion).

EXAMPLE: Which direction is the force on the wire?

ANSWER: 1) Draw in current arrows (+ve to –ve). 2) Fleming's LHR. 3) Draw in direction of force (motion).

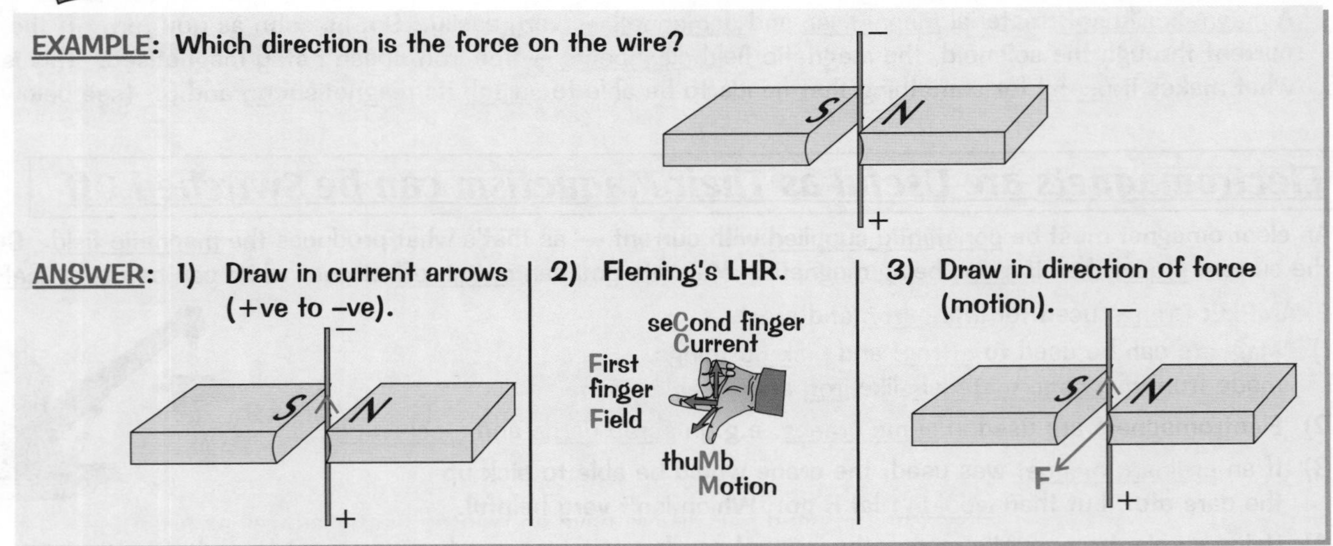

Remember the Left Hand Rule for Motors — drive on the left...

Always remember that it's the LEFT hand rule. If you whip your right hand out in the exam and start looking at the fingers, you'll get it WRONG. Remember that magnetic fields go from north to south, not south to north. And yes, it seems weird that magnets move wires, but that's physics for you.

The Simple Electric Motor

Electric motors use the motor effect (p.82) to get them (and keep them) moving.

The Simple Electric Motor

4 Factors which Speed it up

1) More CURRENT
2) More TURNS on the coil
3) STRONGER MAGNETIC FIELD
4) A SOFT IRON CORE in the coil

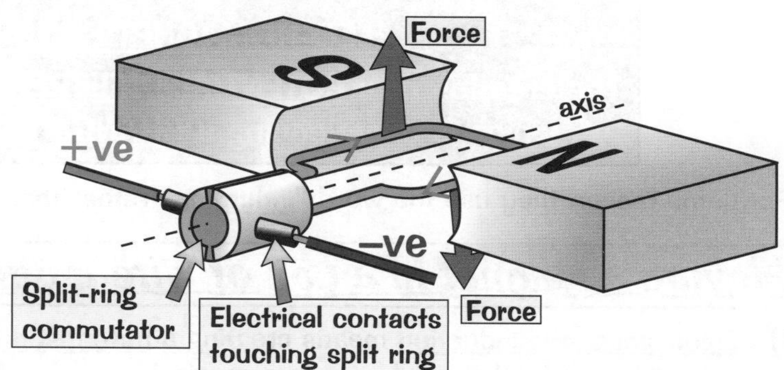

1) The diagram shows the forces acting on the two side arms of the coil of wire.
2) These forces are just the usual forces which act on any current in a magnetic field.
3) Because the coil is on a spindle and the forces act one up and one down, it rotates.
4) The split-ring commutator is a clever way of "swapping the contacts every half turn to keep the motor rotating in the same direction". (Learn that statement because they might ask you.)
5) The direction of the motor can be reversed either by swapping the polarity of the direct current (DC) supply or swapping the magnetic poles over.

Remember, direct current is current that only flows in one direction.

EXAMPLE: Is the coil turning clockwise or anticlockwise?

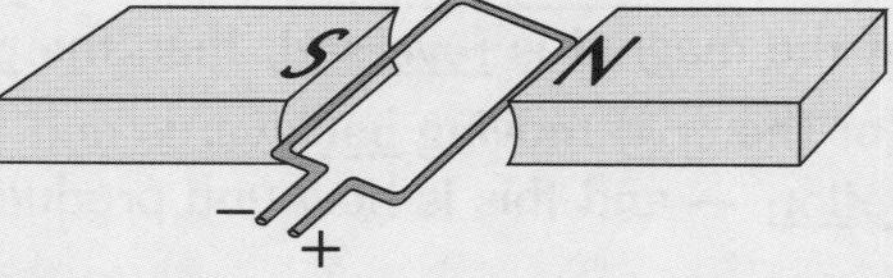

ANSWER: 1) Draw in current arrows (+ve to –ve).

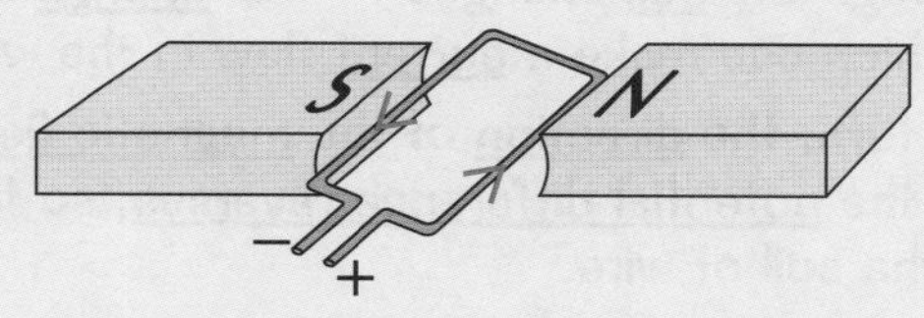

2) Fleming's LHR on one side arm (I've used the right hand arm).

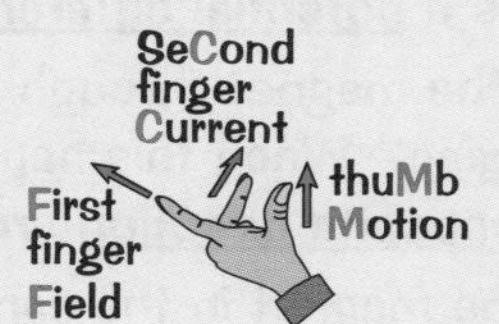

3) Draw in direction of force (motion).

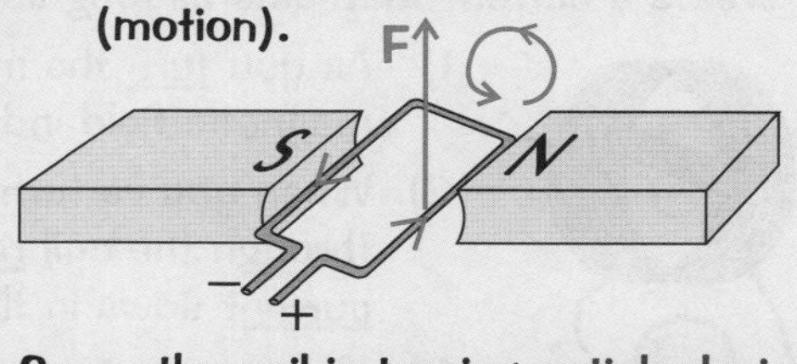

So — the coil is turning anticlockwise.

Electric Motors are used in: CD Players, Food Mixers, Fan Heaters...

...Fans, Printers, Drills, Hair Dryers, Cement Mixers, etc.

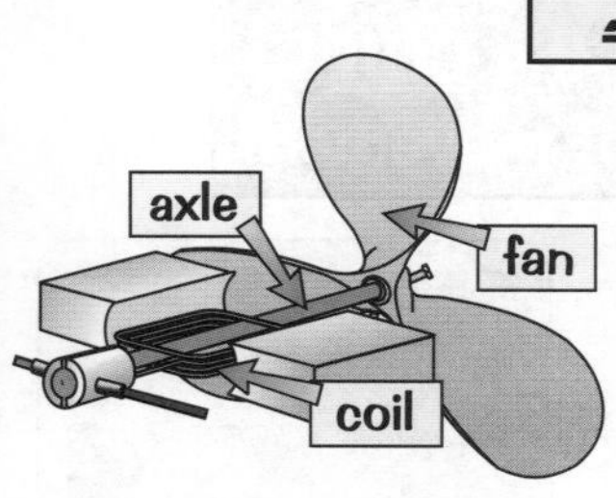

1) Link the coil to an axle, and the axle spins round.
2) In the diagram there's a fan attached to the axle, but you can stick almost anything on a motor axle and make it spin round.
3) For example, in a food mixer, the axle's attached to a blade or whisks. In a CD player the axle's attached to the bit you sit the CD on. Fan heaters and hair dryers have an electric heater as well as a fan.

Hey, don't call my electric motor simple...

Crikey, those electric motors get everywhere. Life'd be a much sadder place without them and my hair would look even more ridiculous. It's all thanks to the wonderful, the splendiferous, the awesome, motor effect.

Electromagnetic Induction

Sounds terrifying. Well, sure it's quite mysterious, but it isn't that complicated:

ELECTROMAGNETIC INDUCTION:
The creation of a POTENTIAL DIFFERENCE across a conductor which is experiencing a CHANGE IN MAGNETIC FIELD.

Remember — potential difference is just another name for voltage.

For some reason they use the word "induction" rather than "creation", but it amounts to the same thing.

Moving a Magnet in a Coil of Wire Induces a Voltage

If the conductor is part of a complete circuit, a current will flow.

1) Electromagnetic induction means creating a potential difference across the ends of a conductor (e.g. a wire).
2) You can do this by moving a magnet in a coil of wire or moving an electrical conductor in a magnetic field ("cutting" magnetic field lines). Shifting the magnet from side to side creates a little "blip" of current.

A few examples of electromagnetic induction:

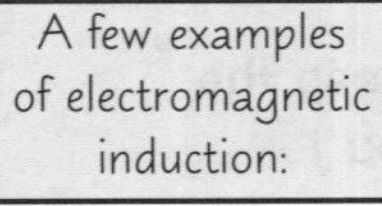
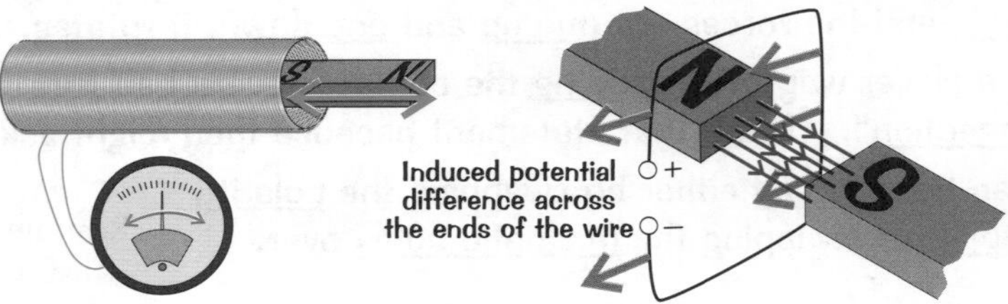

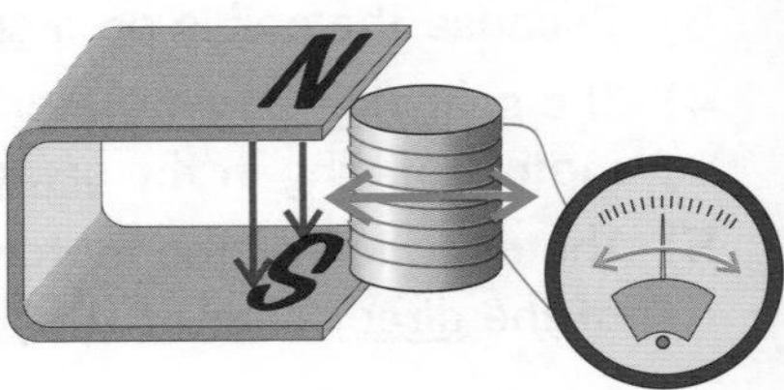

3) If you move the magnet in the opposite direction, then the potential difference/current will be reversed too. Likewise, if the polarity of the magnet is reversed, then the potential difference/current will be reversed too.
4) If you keep the magnet (or the coil) moving backwards and forwards, you produce a potential difference that keeps swapping direction — and this is how you produce an alternating current (AC) — see p.63.

You can create the same effect by turning a magnet end to end in a coil, to create a current that lasts as long as you spin the magnet. This is how generators work.

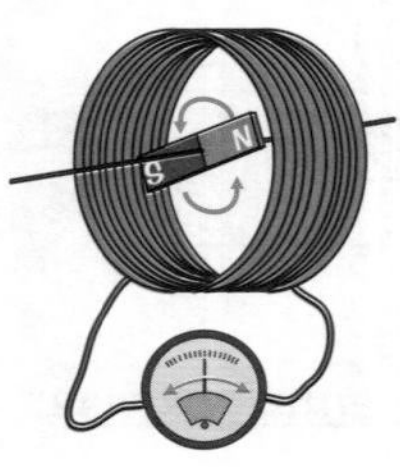

1) As you turn the magnet, the magnetic field through the coil changes — this change in the magnetic field induces a potential difference, which can make a current flow in the wire.
2) When you've turned the magnet through half a turn, the direction of the magnetic field through the coil reverses. When this happens, the potential difference reverses, so the current flows in the opposite direction around the coil of wire.
3) If you keep turning the magnet in the same direction — always clockwise, say — then the potential difference will keep on reversing every half turn and you'll get an AC current.

Four Factors Affect the Size of the Induced Voltage

1) If you want a different peak voltage (and current) you have to change the rate that the magnetic field is changing. For a bigger voltage you need to increase at least one of these four things:

1) The STRENGTH of the MAGNET	2) The AREA of the COIL
3) The number of TURNS on the COIL	4) The SPEED of movement

2) To reduce the voltage, you would reduce one of those factors, obviously.
3) If you turn the magnet faster, you'll get a higher peak voltage, but also a higher frequency — because the magnetic field is reversing more frequently.

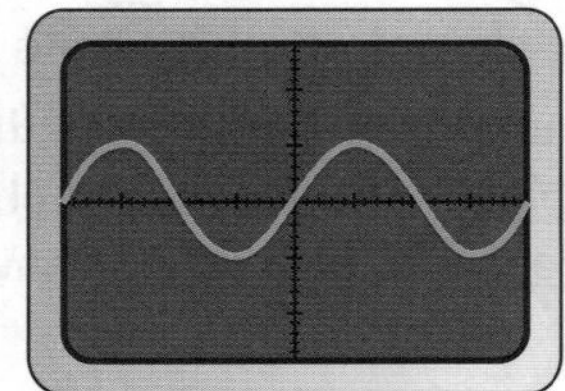

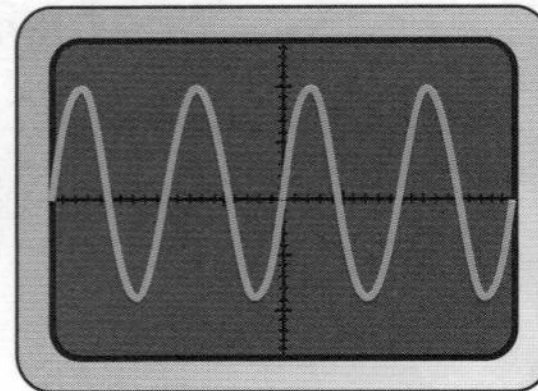

EM induction — works whether the coil or the field is moving...

Tricky stuff — but useful. This is how most of our electricity is generated, whether it's in a coal-fired power station or a wind turbine. Steam, or wind, or whatever turns a turbine attached to a coil inside a magnetic field.

Transformers

Transformers use electromagnetic induction to change potential difference (p.d.). So they will only work on AC.

Transformers Change the p.d. — but only AC p.d.

There are a few different types of transformer. The two you need to know about are step-up transformers and step-down transformers. They both have two coils, the primary and the secondary, joined with an iron core.

STEP-UP TRANSFORMERS step the voltage up. They have more turns on the secondary coil than the primary coil.

Iron core
Magnetic field in the iron core
Primary coil
Secondary coil

STEP-DOWN TRANSFORMERS step the voltage down. They have more turns on the primary coil than the secondary.

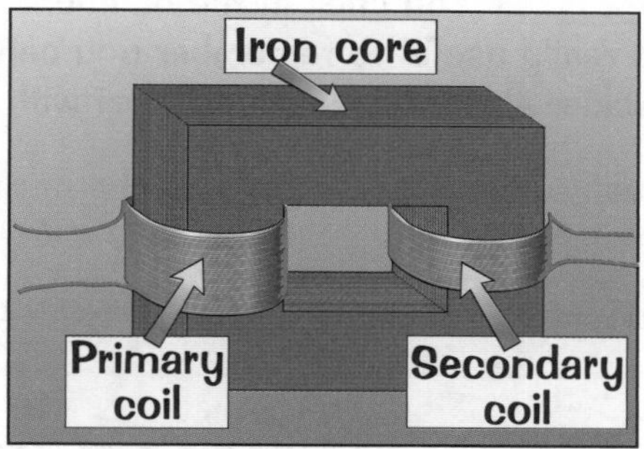

The iron core is purely for transferring the changing magnetic field from the primary coil to the secondary (see below). No electricity flows round the iron core.

Transformers Work by Electromagnetic Induction

1) The primary coil produces a magnetic field which stays within the iron core. This means nearly all of it passes through the secondary coil and hardly any is lost.
2) Because there is alternating current (AC) in the primary coil, the field in the iron core is constantly changing direction (100 times a second if it's at 50 Hz) — i.e. it is a changing magnetic field.
3) This rapidly changing magnetic field is then felt by the secondary coil.
4) The changing field induces an alternating potential difference across the secondary coil (with the same frequency as the alternating current in the primary) — electromagnetic induction of a potential difference in fact.
5) The relative number of turns on the two coils determines whether the potential difference induced in the secondary coil is greater or less than the potential difference in the primary.
6) In a step-up transformer, the p.d. across the secondary coil is greater than the p.d. across the primary coil.
7) In a step-down transformer, the p.d. across the secondary coil is less than the p.d. across the primary coil.
8) If you supplied DC to the primary, you'd get nothing out of the secondary at all. Sure, there'd still be a magnetic field in the iron core, but it wouldn't be constantly changing, so there'd be no induction in the secondary because you need a changing field to induce a potential difference. Don't you! So don't forget it — transformers only work with AC. They won't work with DC at all.

Transformers are Nearly 100% Efficient So "Power In = Power Out"

The formula for power supplied is: Power = Current × Potential Difference or: $P = I \times V$ (see page 67).

So you can write electrical power input = electrical power output as:

$$V_pI_p = V_sI_s$$

V_p = p.d. across primary coil (V)
I_p = current in the primary coil (A)
V_s = p.d. across secondary coil (V)
I_s = current in the secondary coil (A)

EXAMPLE: A transformer in a travel adaptor steps up a 110 V AC mains electricity supply to the 230 V needed for a hair dryer. The current through the hair dryer is 5 A. If the transformer is 100% efficient, calculate how much current is drawn by the transformer from the mains supply.

ANSWER: $V_p \times I_p = V_s \times I_s$ so $110 \times I_p = 230 \times 5$ $I_p = (230 \times 5) \div 110 = 10.5$ A

Transformers

Ah, more about transformers. And as per usual, some equations too. I don't like change.

The Transformer Equation — use it Either Way Up

$$\frac{V_P}{V_S} = \frac{n_P}{n_S} \quad \text{or} \quad \frac{V_S}{V_P} = \frac{n_S}{n_P}$$

You can calculate the output potential difference from a transformer if you know the input potential difference and the number of turns on each coil.

$$\frac{\text{Potential Difference across Primary Coil}}{\text{Potential Difference across Secondary Coil}} = \frac{\text{Number of turns on Primary Coil}}{\text{Number of turns on Secondary Coil}}$$

Well, it's just another formula. You stick in the numbers you've got and work out the one that's left. It's really useful to remember you can write it either way up — this example's much trickier algebra-wise if you start with V_s on the bottom...

> EXAMPLE: A transformer has 40 turns on the primary and 800 on the secondary. If the input potential difference is 1000 V, find the output potential difference.
>
> ANSWER: $V_S/V_P = n_S/n_P$ so $V_S/1000 = 800/40$ $V_S = 1000 \times (800/40) =$ 20 000 V

Switch Mode Transformers are used in Chargers and Power Supplies

See page 85 for more on frequency and transformers.

1) Switch mode transformers are a type of transformer that operates at higher frequencies than traditional transformers — usually between 50 kHz and 200 kHz.
2) Because they work at higher frequencies, they can be made much lighter and smaller than traditional transformers that work from a 50 Hz mains supply.
3) This makes them more useful in things like mobile phone chargers and power supplies, e.g. for laptops.
4) Switch mode transformers are more efficient than other types of transformer. They use very little power when they're switched on but no load (the thing you're charging or powering) is applied, e.g. if you've left your phone charger plugged in but haven't attached your phone.

Isolating Transformers are Used in Bathrooms

1) Isolating transformers have an equal number of turns in the primary and secondary coils, so also have equal primary and secondary voltages. They can be found in some mains circuits in the home, such as in a bathroom shaver socket.
2) The isolating transformer inside the shaver socket allows you to use the shaver without being physically connected to the mains. So it minimises your risk of getting electrocuted. Phew.

Transformers Are Used on the National Grid

See page 25 for more on the National Grid.

1) To transmit a lot of power, you either need high voltage or high current (P = VI).
2) The problem with high current is the loss (as heat) due to the resistance of the cables (and transformers).
3) The formula for power loss due to resistance in the cables is:

$$\text{Power Loss} = \text{Current}^2 \times \text{Resistance}$$

$$P = I^2R$$

4) Because of the I^2 bit, if the current is 10 times bigger, the losses will be 100 times bigger. So it's much cheaper to boost the voltage up to 400 000 V and keep the current very low.
5) The transformers have to step the voltage up at one end, for efficient transmission, and then bring it back down to safe, usable levels at the other end.

Which transformer do you need to enslave the Universe — Megatron...

You'll need to practise with those tricky equations. The transformer equation's unusual because it can't be put into a formula triangle, but other than that the method is the same — stick in the numbers. Just practise.

Revision Summary for Section Six

There's only one way to check you know it all. Sorry.

1) Give a definition of a magnetic field.
2) Sketch magnetic fields for: a) a current-carrying wire, b) a rectangular coil, c) a solenoid.
3) What is an electromagnet? Describe one use of electromagnets. Explain why they're good for this job.
4) Describe what happens to a current-carrying wire when it is placed in a magnetic field. What is the name of this effect?
5) Describe the three details of Fleming's left hand rule. What is it used for?
6) The diagrams show a simple electric motor. The coil is turning clockwise. Which diagram, A or B, shows the correct polarity of the magnets?

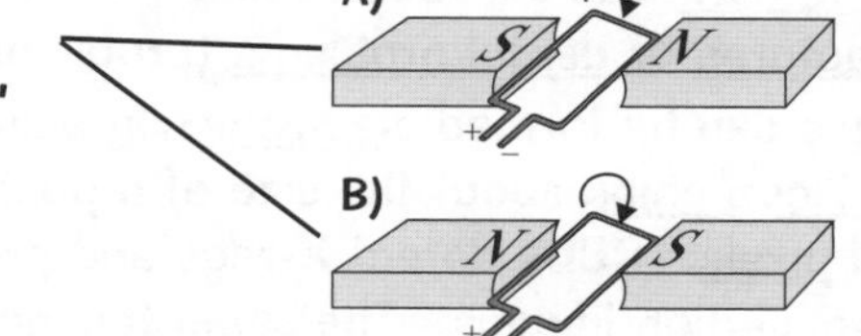

7) List four ways to speed up a motor.
8) Give the definition of electromagnetic induction.
9) Describe two ways in which you could induce a voltage using a wire and a magnet.
10) List four factors which affect the size of an induced voltage.
11) Sketch two types of transformer and explain the differences between them.
12) An engineering executive is travelling from the USA to Italy and taking a computer monitor with him. In the USA, domestic electricity is 110 V AC, and in Italy it's 230 V AC. What kind of transformer would the engineering executive need to plug his monitor into?
13) Explain how a transformer works and why transformers only work on AC voltage.
14)* A transformer steps down 230 V from the mains supply to the 130 V needed for an appliance. If the transformer draws 2 A from the mains supply, calculate how much current goes through the appliance.
15) Write down the transformer equation.
16)* A transformer has 20 turns on the primary coil and 600 on the secondary coil. If the input potential difference is 9 V, find the output potential difference.
17)* A transformer has an input voltage of 20 V and an output voltage of 16 V. If there are 64 turns on the secondary coil, how many turns are there on the primary coil?
18) Give two advantages of switch mode transformers over traditional transformers.
19) Write down three facts about isolating transformers.
20) Explain why power in power lines is transmitted at such a high voltage.
21)* Calculate how much power is lost by 8 A of current flowing through a cable with resistance of 22 Ω.

* Answers on page 140.

X-rays in Medicine

X-rays are ionising — they can damage living cells (see p.35) but they can be really useful if handled carefully...

X-ray Images are Used in Hospitals for Medical Diagnosis

1) X-rays are high frequency, short wavelength electromagnetic waves (see p.32). Their wavelength is roughly the same size as the diameter of an atom.
2) They are transmitted by (pass through) healthy tissue, but are absorbed by denser materials like bones and metal.
3) They affect photographic film in the same way as light, which means they can be used to take photographs.
4) X-ray photographs can be used to diagnose many medical conditions such as bone fractures or dental problems (problems with your teeth).
5) X-ray images can be formed electronically using charge-coupled devices (CCDs). CCDs are silicon chips about the size of a postage stamp, divided up into a grid of millions of identical pixels. CCDs detect X-rays and produce electronic signals which are used to form high resolution images. The same technology is used to take photographs in digital cameras.

CT Scans use X-rays

Computerised axial tomography (CT) scans use X-rays to produce high resolution images of soft and hard tissue.

The patient is put inside the cylindrical scanner, and an X-ray beam is fired through the body from an X-ray tube and picked up by detectors on the opposite side. The X-ray tube and detectors are rotated during the scan. A computer interprets the signals from the detectors to form an image of a two-dimensional slice through the body. Multiple two-dimensional CT scans can be put together to make a three-dimensional image of the inside of the body.

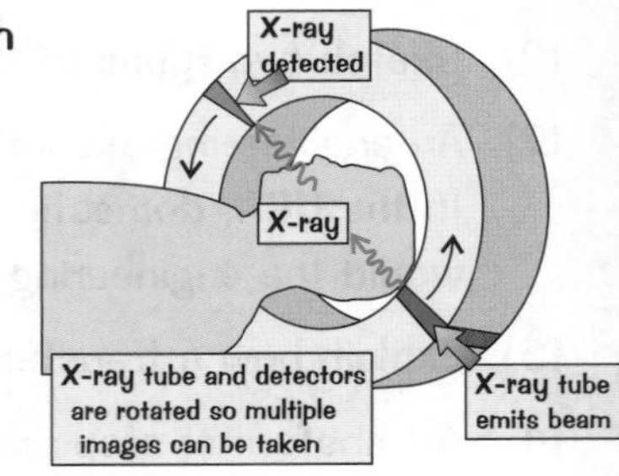

Soft tissue can absorb a small amount of X-ray radiation. CT scans use lots of X-rays (more than normal X-ray photographs) to distinguish between the tiny variations in tissue density.

X-rays can be Used to Treat Cancer

X-rays can cause ionisation — high doses of X-rays will kill living cells. They can therefore be used to treat cancers, just like gamma radiation (see p.76).

1) The X-rays are focused on the tumour using a wide beam.
2) This beam is rotated round the patient with the tumour at the centre.
3) This minimises the exposure of normal cells to radiation, and so reduces the chances of damaging the rest of the body.

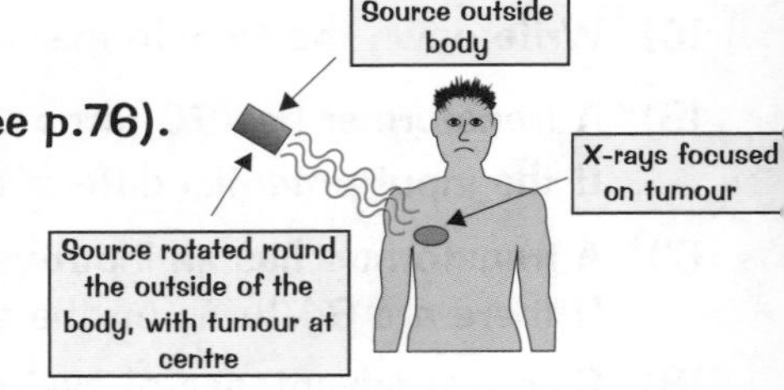

Fluoroscopes use X-rays to Create Moving Images of Patients' Insides

1) Basic fluoroscopy works by placing a patient between an X-ray source and a fluorescent screen. The intensity of X-rays reaching the screen will vary depending on what they've passed through in the body.
2) The fluorescent screen absorbs the X-rays and fluoresces (gives off light) to show a live image on the screen. The higher the intensity of the X-rays, the brighter that bit of the screen.
3) In modern fluoroscopy, an image intensifier is used to increase the brightness of the image (up to 5000 times brighter). This means a lower dose of X-rays can be given to the patient.
4) Fluoroscopy is used to diagnose problems in the way organs are functioning. Because X-rays pass easily through soft tissue, the patient is given a 'contrast medium'. This is a substance which improves the contrast of the image seen by 'enhancing' the soft tissue — making it more visible.

Don't just scan this page — focus on it...

So, traditional X-ray photographs are only useful when looking at hard materials like bone, but CT scans and fluoroscopes use X-rays to take pictures of the squidgy stuff as well as the hard stuff. X-raytastic.

Ultrasound

There's sound, and then there's ultrasound.

Ultrasound is Sound with a Higher Frequency than We Can Hear

Electrical systems can be made which produce electrical oscillations of any frequency.
These can easily be converted into mechanical vibrations to produce sound waves of a higher frequency than the upper limit of human hearing (the range of human hearing is 20 to 20 000 Hz). This is called ultrasound.

Ultrasound Waves Get Partially Reflected at a Boundary Between Media

1) When a wave passes from one medium into another, some of the wave is reflected off the boundary between the two media, and some is transmitted (and refracted). This is partial reflection.
2) What this means is that you can point a pulse of ultrasound at an object, and wherever there are boundaries between one substance and another, some of the ultrasound gets reflected back.
3) The time it takes for the reflections to reach a detector can be used to measure how far away the boundary is.
4) This is how ultrasound imaging works (see p.90).

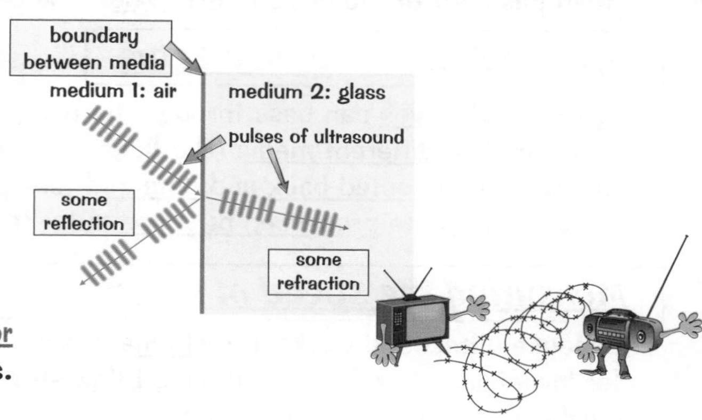

You Can Use Oscilloscope Traces to Find Boundaries

1) The oscilloscope trace on the right shows an ultrasound pulse reflecting off two separate boundaries.
2) Given the "seconds per division" setting of the oscilloscope (see p.63), you can work out the time between the pulses by measuring on the screen.
3) If you know the speed of sound in the medium, you can work out the distance between the boundaries, using this formula:
s is distance in metres, m.
v is speed in metres per second, m/s.
t is time in seconds, s.

$$s = v \times t$$

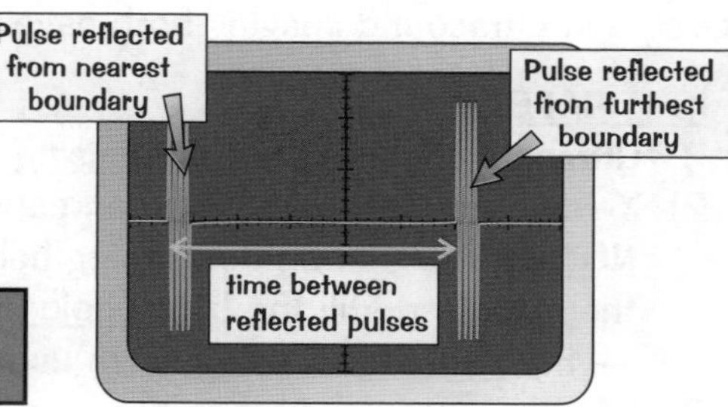

Example: A pulse of ultrasound is beamed into a patient's abdomen. The first boundary it reflects off is between fat and muscle. The second boundary is between muscle and a body cavity. An oscilloscope trace shows that the time between the reflected pulses is 10 μs. The ultrasound travels at a speed of 1500 m/s. Calculate the distance between the fat/muscle boundary and the muscle/cavity boundary.

1 μs = 0.000001 s

So, you'll need to find the distance using $s = v \times t$.
BUT, the reflected pulses have travelled there and back, so the distance you calculate will be twice the distance between boundaries (think about it).

Pulse sent

2 pulses back

$s = v \times t = 1500 \times 0.00001 = 0.015$ m.
So the distance between boundaries $= 0.015 \div 2 = 0.0075$ m = 7.5 mm.

Partially reflected — completely revised...

It's crazy to think that you can use sound waves to make a image — but that's the basis of ultrasound scanning (see next page). And it all comes down to some simple reflection and a bit of distance = speed × time.

Medical Imaging

Those high frequency sound waves are really useful, especially when people are ill or pregnant.

Ultrasound Waves can be Used in Medicine

Ultrasound has a variety of uses in medicine, from investigating blood flow in organs, to diagnosing heart problems, to checking on fetal development. The examples below are three of the most common. Read on...

Breaking Down Kidney Stones

Kidney stones are hard masses that can block the urinary tract — ouch. An ultrasound beam concentrates high-energy waves at the kidney stone and turns it into sand-like particles. These particles then pass out of the body in the urine. The patient doesn't need surgery and it's relatively painless.

Pre-Natal Scanning of a Fetus

Ultrasound waves can pass through the body, but whenever they reach a boundary between two different media (like fluid in the womb and the skin of the fetus) some of the wave is reflected back and detected (see p.89). The exact timing and distribution of these echoes are processed by a computer to produce a video image of the fetus.

Measuring the Speed of Blood Flow

Because ultrasound works in real time it can show things changing and moving. This makes it useful for investigating problems with blood flow — particularly in the heart and liver. Special ultrasound machines can measure the speed of blood flow and identify any blockages in the veins and arteries.

Medical Imaging is Full of Compromises...

Doctors have to make compromises between getting a good enough image to be able to diagnose problems accurately, whilst putting the patient at as low a risk as possible. X-ray and ultrasound imaging both have their advantages and disadvantages...

IS IT SAFE?

1) Ultrasound waves are non-ionising and, as far as anyone can tell, safe.
2) X-rays are ionising. They can cause cancer, and are definitely NOT safe to use on developing babies. Despite the dangers, they're often still the best choice to treat or diagnose a patient — the benefits of using them usually outweigh the risks.
3) CT scans use a lot more X-ray radiation than standard X-ray photographs, so the patient is exposed to even more ionising radiation. Generally CT scans aren't taken unless they are really needed because of the increased radiation dose.

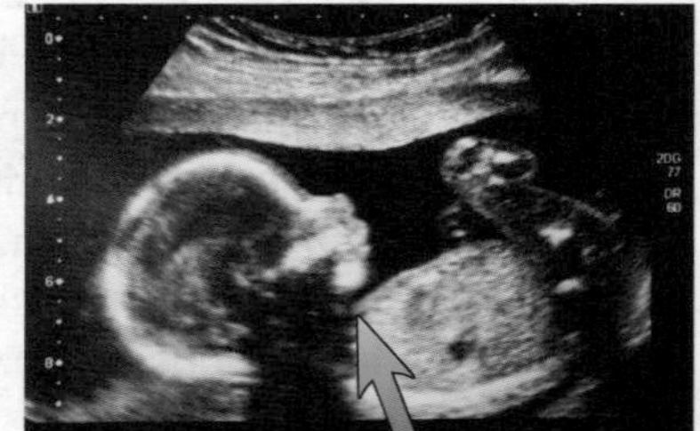

Ultrasound scans are safe for the fetus, but they do give a fuzzy image.

WHAT ABOUT IMAGE QUALITY?

1) Ultrasound images are typically fuzzy — which can make it harder to diagnose some conditions using these images.
2) X-ray photographs produce clear images of bones and metal, but not a lot else.
3) CT scans produce detailed images and can be used to diagnose complicated illnesses, as the high resolution images can make it easier to work out the problem. High quality 3D images can also be used in the planning of complicated surgery.

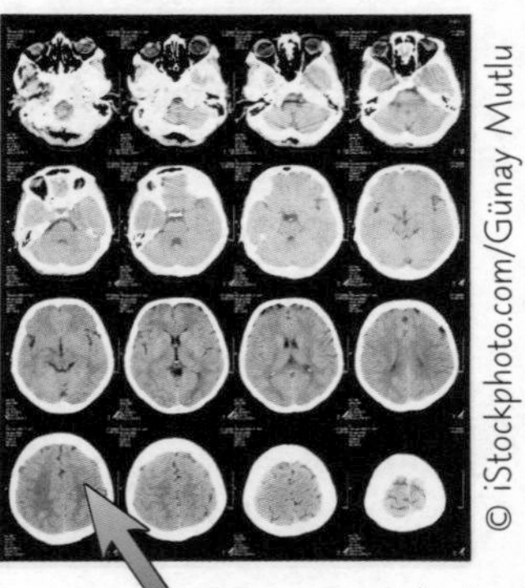

CT scans of the brain are very detailed and clear.

What did the X-ray say to the muscle? Just passing through...

It's important that you're able to compare and contrast the different imaging techniques. Sit down with a pen and paper and write yourself a mini essay all about medical imaging and all the advantages and disadvantages.

Other Uses of Physics in Medicine

Medical physics isn't all X-rays and ultrasound, physics is used in loads of other ways in medicine too.

Endoscopes Use Bundles of Optical Fibres

1) An endoscope is a thin tube containing optical fibres (p.103) that lets surgeons examine inside the body.
2) Endoscopes consist of two bundles of optical fibres — one to carry light to the area of interest and one to carry an image back so that it can be viewed.

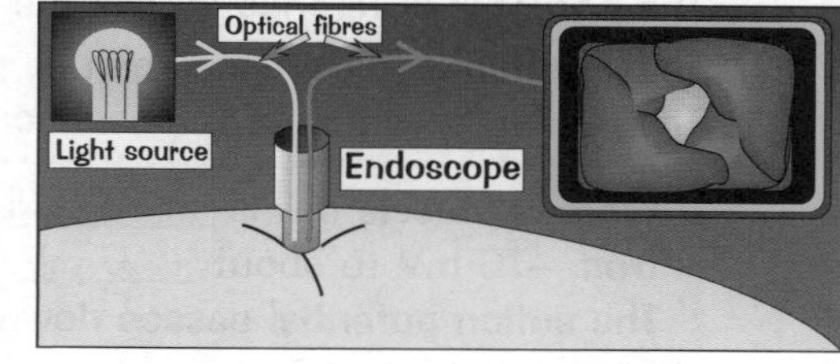

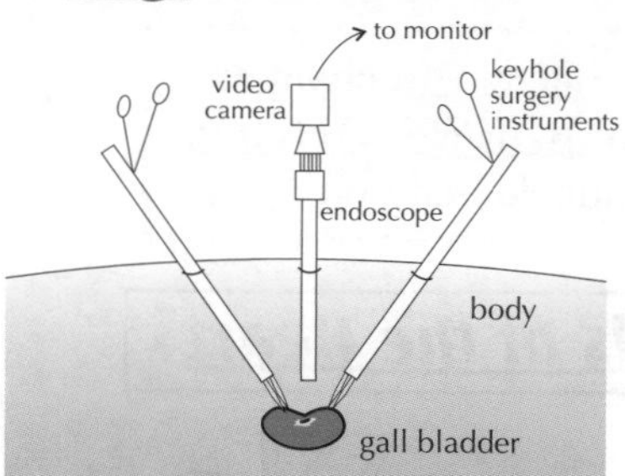

3) The image can be seen through an eyepiece or displayed as a full-colour moving image on a TV screen.
4) The big advantage of using endoscopes is that surgeons can now perform many operations by only cutting teeny holes in people — this is called keyhole surgery, and it wasn't possible before optical fibres.

Pulse Oximeters Use Light to Check the % Oxygen in the Blood

Pulse oximetry measures the amount of oxygen carried by the haemoglobin in a patient's blood.

- Haemoglobin carries oxygen around your body from your lungs to your cells.
- Haemoglobin changes colour depending on its oxygen content. Oxygen-rich haemoglobin (oxyhaemoglobin) is bright red. After giving up its oxygen to cells, it appears purply.

1) A pulse oximeter has a transmitter, which emits two beams of red light. It also has a photo detector to measure light.
2) These are placed either side of a thin part of the body, e.g. a finger.
3) As the beams of light pass through the tissue, some of the light is absorbed by the blood.
4) The amount of light absorption depends on the colour of the blood, which depends on its oxyhaemoglobin content. Healthy people normally have at least 95% oxyhaemoglobin in their arteries.
5) Reflection pulse oximetry is similar, but it reflects light off red blood cells instead.

emitters of red and infrared light
detector
display
98%

PET Scanning Involves Positron/Electron Annihilation

1) A positron is the antiparticle of an electron — it has the same mass but opposite charge.
2) When a particle meets its antiparticle (its opposite), the result is annihilation. All the mass of both particles is converted into energy, which is given off in the form of gamma rays.
3) Einstein said that mass is a form of energy — you can think of them as being the same thing, mass energy.
4) Mass energy is conserved in annihilation reactions. It's Einstein's famous equation at work:

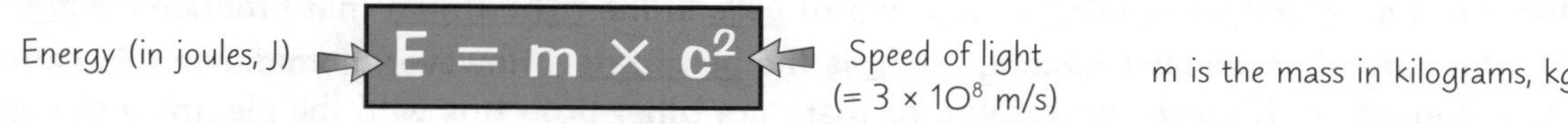

Energy (in joules, J) $E = m \times c^2$ Speed of light ($= 3 \times 10^8$ m/s)   m is the mass in kilograms, kg

5) In PET (positron emission tomography) scanning, a positron-emitting radioisotope is injected into a patient. The emitted positrons collide with electrons in the organs, causing them to annihilate and emit high-energy gamma rays. There will be a higher take-up of the radioisotope in tumour cells than in normal cells.
6) Detectors around the body detect each pair of gamma rays. By detecting at least three pairs, the location of the tumour can be accurately found.

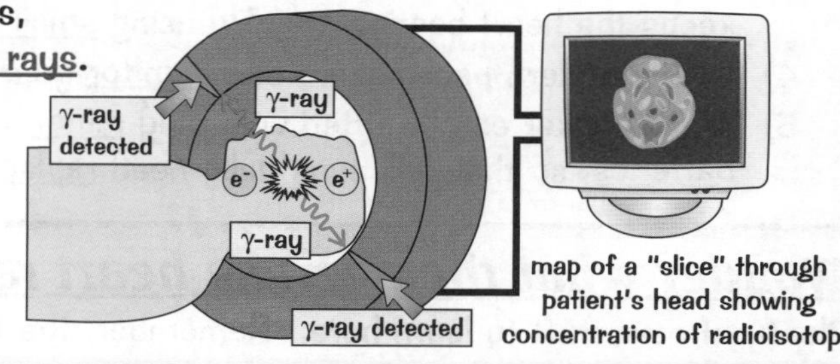

Other Uses of Physics in Medicine

Ah, yet one more glorious page on how the awesome power of physics is wielded for the good of medicine.

Muscle Cells Can Generate Potential Differences

1) Between the inside of a muscle cell and the outside, there's a potential difference (a voltage). The potential difference across the cell membrane of a muscle cell at rest is called the resting potential.
2) These potential differences can be measured with really teeny tiny needle electrodes. The resting potential of a muscle cell is about –70 mV (millivolts).

When a muscle cell is stimulated by an electrical signal, the potential difference changes from –70 mV to about +40 mV. This increased potential is called an ACTION POTENTIAL. The action potential passes down the length of the cell, making the muscle cell contract.

Electrocardiographs Measure the Action Potentials of the Heart

1) The heart is a pump made of muscle, which is split up into four chambers — the atria at the top and the ventricles at the bottom.
2) When the heart beats, an action potential passes through the atria, making them contract. A fraction of a second later, another action potential passes through the ventricles, making them contract too.
3) Once the action potential has passed, the muscle relaxes.
4) These action potentials produce weak electrical signals on the skin.
5) An electrocardiograph records the action potentials of the heart using electrodes stuck onto the chest, arms and legs. For accurate readings the patient should lie or sit still and relax.
6) The results are displayed on a screen or printed out as a graph called an electrocardiogram (ECG), and are used to look at the action of the heart.

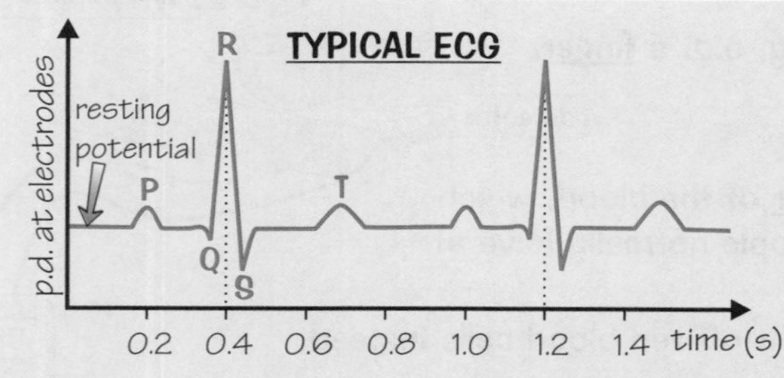

LEARN the basic shape and what it means:

- The horizontal line is just the resting potential.
- The 'blip' at P shows the contraction of the atria.
- The QRS blip shows the contraction of the ventricles. It's a weird shape because you've got the relaxation of the atria going on there too.
- And T shows the relaxation of the ventricles.

7) You can work out the heart rate from an ECG using:

$$\text{frequency (hertz)} = \frac{1}{\text{time period (seconds)}}$$

On the graph, the time from peak to peak is 0.8 s. So frequency = 1/0.8 = 1.25 Hz. Multiplying by 60 converts this into a heart rate in beats per minute: 1.25 × 60 = 75 beats per minute.

A Pacemaker is a Device used to Regulate Heart Beat

1) The heart has a natural pacemaker — a group of cells in the right atrium that produce electrical signals.
2) The heart's natural pacemaker directly controls the heart rate. However, sometimes this natural pacemaker isn't fast enough, or it pulses irregularly, or there are other problems with the electrical signals being sent.
3) People with these kinds of problems may be fitted with an artificial pacemaker — a device that keeps the heart beating steadily using small electric impulses to stimulate the heart to beat.
4) Some modern pacemakers can monitor your breathing, and adjust your heart rate to match your activity.
5) A pacemaker can be fitted with only minor surgery. However, artificial pacemakers are powered by batteries, so they will eventually need replacing when the battery loses power.

Wonder what the average heart rate is during exams...

Plenty of new stuff to learn here. Remember, the time period (T) is the time from peak to peak. Good-o.

Revision Summary for Section Seven

Another section conquered, you absolute legend. Now all you need to do is just answer these questions to see how much you've learnt. I bet it's loads.

1) What are X-rays?
2) Name two materials X-rays are absorbed by.
3) What is a charge-coupled device?
4) How do CT scans form images of the body?
5) Describe how X-rays can be used to treat cancer.
6) Describe how a fluoroscope can be used to create images of the inside of a patient's body.
7) What is ultrasound?
8) Describe three different ways ultrasound is used in medicine.
9)* Ultrasound travels through fat at a speed of 1000 m/s.
A pulse of ultrasound is sent into a person and is partially reflected off a layer of fat and a layer of muscle. The time between two reflected pulses of ultrasound is 0.00004 s. How thick is the layer of fat?
10) Explain why ultrasound rather than X-rays are used to take images of a fetus.
11) What are the advantages of using CT scans over ultrasound scans?
12) Suggest why X-rays are still used to scan patients despite the risks from using ionising radiation.
13) a) What is an endoscope?
b) Name one medical technique made possible by endoscopy.
14) Describe how pulse oximetry works.
15) What particle is the antiparticle of an electron?
16) Name the reaction that happens when a particle and its antiparticle collide.
17) Write down Einstein's famous equation that relates mass and energy.
18) What is 'PET' scanning short for?
19) PET scans can detect cancer tumours. Briefly explain how.
20) What's meant by the 'resting potential' of a muscle cell?
21) What's meant by an 'action potential'?
22) Sketch the basic shape of an ECG trace.
23)*An ECG shows that Karen's heart sends out a strong electrical signal every 0.7 seconds.
Calculate Karen's heart rate in beats per minute.
24) What are artificial pacemakers used for?

*Answers on page 140.

Turning Forces and the Centre of Mass

Moments, they're magic. Or maybe not. Either way, expect to be royally sick of pivots by the end of the page.

A Moment is the Turning Effect of a Force

The size of the moment of the force is given by:

MOMENT = FORCE × perpendicular DISTANCE from the line of action of the force to the pivot

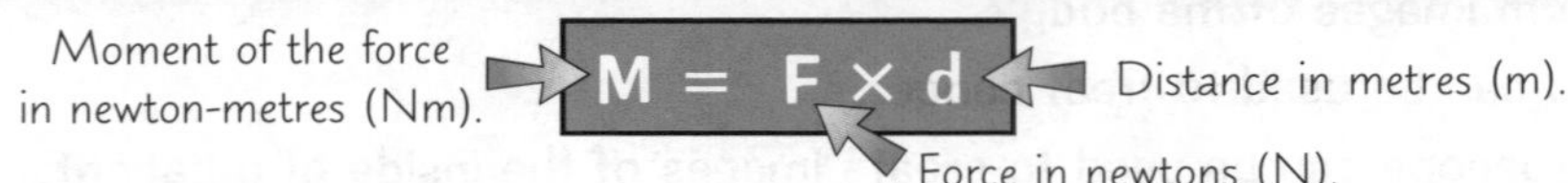

1) The force on the spanner causes a turning effect or moment on the nut (which acts as a pivot). A larger force would mean a larger moment.

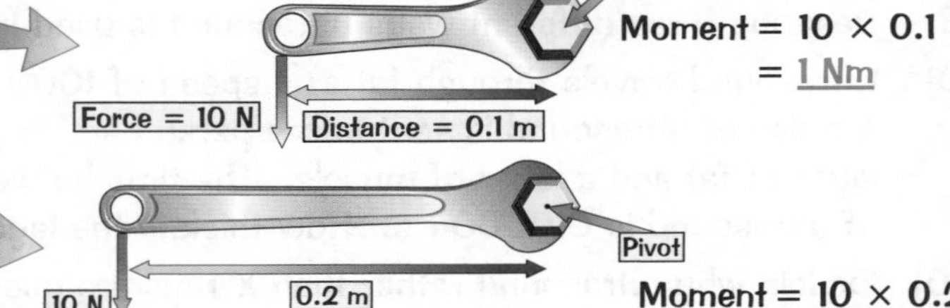

2) Using a longer spanner, the same force can exert a larger moment because the distance from the pivot is greater (see p.95).

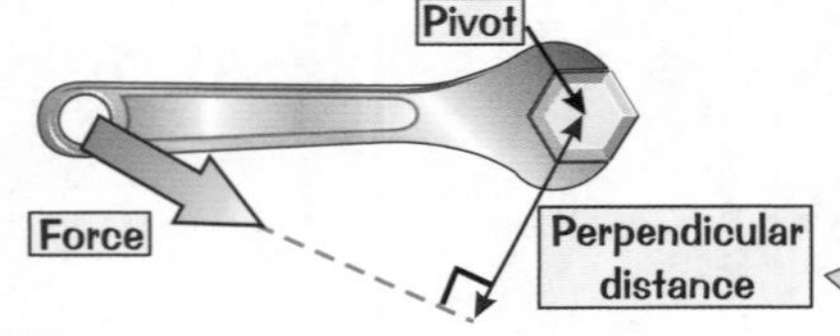

3) To get the maximum moment (or turning effect) you need to push at right angles (perpendicular) to the spanner.
4) Pushing at any other angle means a smaller moment because the perpendicular distance between the line of action and the pivot is smaller.

The Centre of Mass Hangs Directly Below the Point of Suspension

1) You can think of the centre of mass of an object as the point at which the whole mass is concentrated.
2) A freely suspended object will swing until its centre of mass is vertically below the point of suspension.

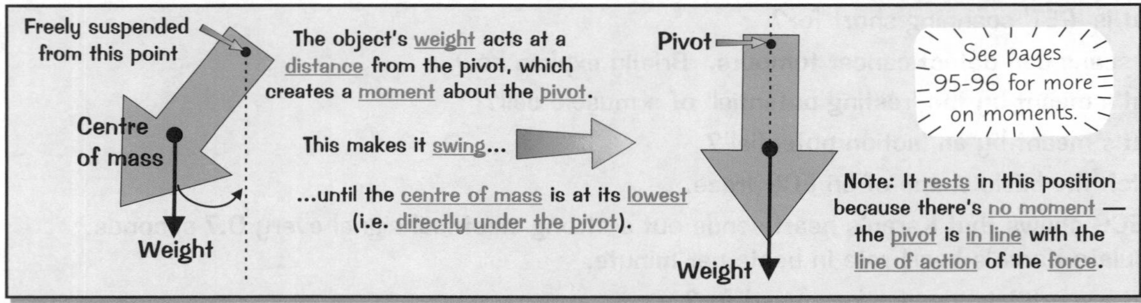

3) This means you can find the centre of mass of any flat shape like this:

a) Suspend the shape and a plumb line from the same point, and wait until they stop moving.
b) Draw a line along the plumb line.
c) Do the same thing again, but suspend the shape from a different pivot point.
d) The centre of mass is where your two lines cross.

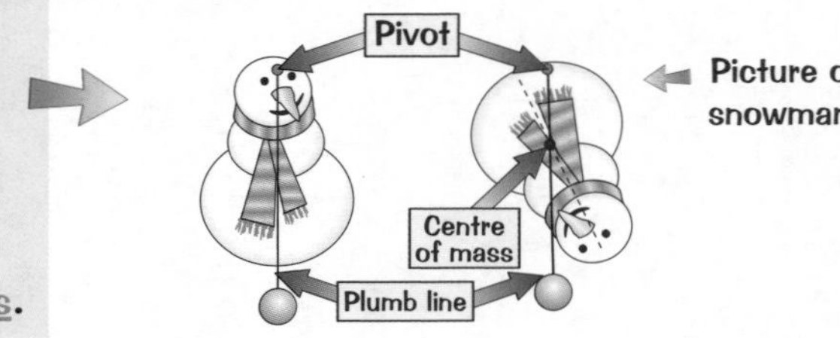

4) But you don't need to go to all that trouble for symmetrical shapes. You can quickly guess where the centre of mass is by looking for lines of symmetry.

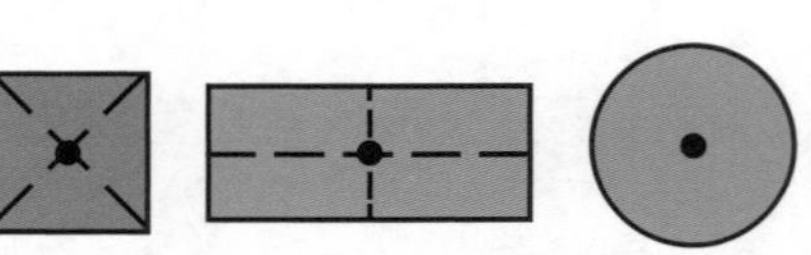

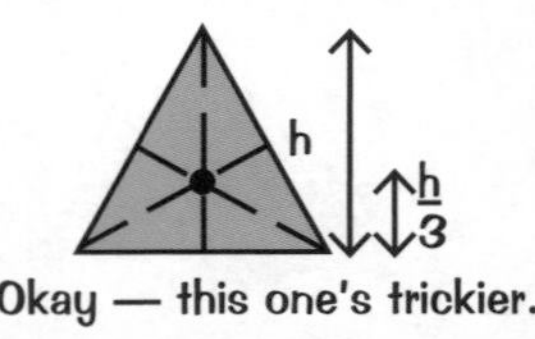

Be at the centre of mass — sit on the middle pew...

So there you go, how to find the centre of mass of your favourite piece of irregularly-shaped paper in a few easy steps. You should also now know that the next time someone asks you "How's it hanging?" your response should be "Directly below the point of suspension, thank you for asking". This page truly was an education.

Balanced Moments and Levers

Once you can calculate moments, you can work out if a seesaw is balanced. Useful thing, physics.

A Question of Balance — Are the Moments Equal?

If the anticlockwise moments are equal to the clockwise moments, the object won't turn.

Example 1: Your younger brother weighs 300 N and sits 2 m from the pivot of a seesaw. If you weigh 700 N, where should you sit to balance the seesaw?

For the seesaw to balance: **Total Anticlockwise Moments = Total Clockwise Moments**

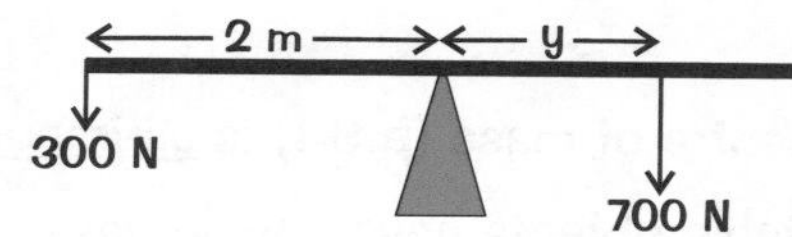

anticlockwise moment = clockwise moment
$300 \times 2 = 700 \times y$
$y = 0.86$ m

Ignore the weight of the seesaw — its centre of mass is on the pivot, so it doesn't have a turning effect.

Example 2: A 6 m long steel girder weighing 1000 N rests horizontally on a pole 1 m from one end. What is the tension in a supporting cable attached vertically to the other end?

The 'tension in the cable' bit makes it sound harder than it actually is.
But the girder's weight is balanced by the tension force in the cable, so...

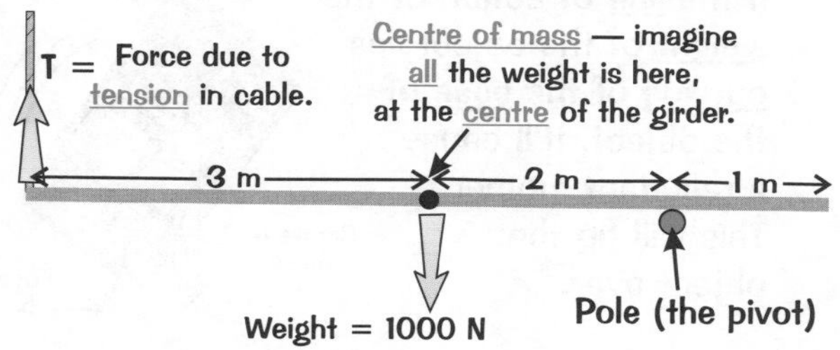

anticlockwise moment (due to weight) = clockwise moment (due to tension in cable)
$1000 \times 2 = T \times 5$
$2000 = 5T$
and so $T = 400$ N

Simple Levers use Balanced Moments

Levers use the idea of balanced moments to make it easier for us to do work (e.g. lift an object):

1) The moment needed to do work = force x distance from the pivot (see previous page). So the amount of force needed to do work depends on the distance the force is applied from the pivot.
2) Levers increase the distance from the pivot at which the force is applied — so this means less force is needed to get the same moment.
3) That's why levers are known as force multipliers — they reduce the amount of force that's needed to get the same moment by increasing the distance.

Examples of Simple Levers as Force Multipliers

Scissors use a combination of two levers.

Long sticks or bars:

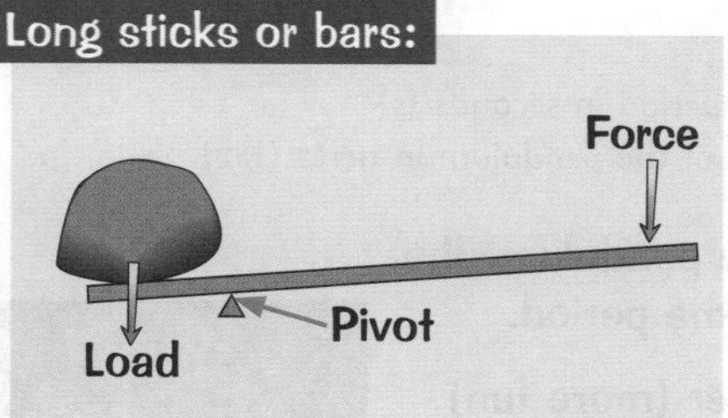

Wheelbarrows:

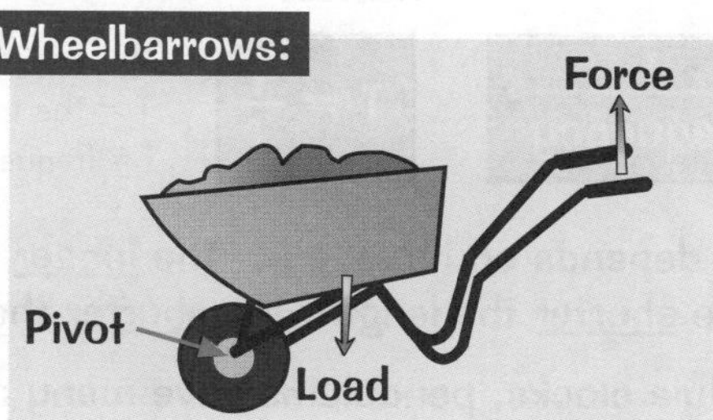

Scissors:

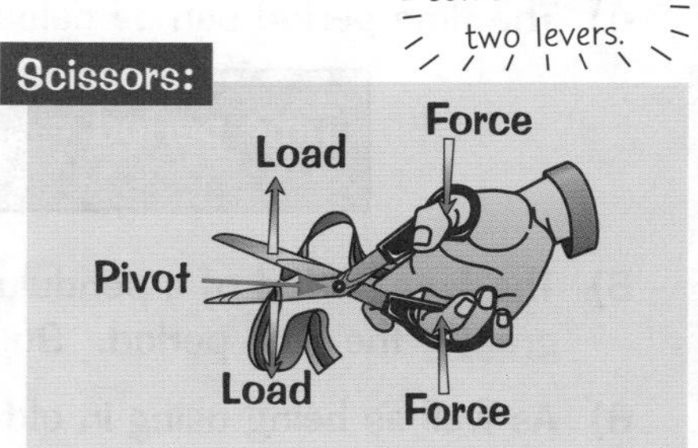

These levers make it easier to do work by moving the distance the force is applied further from the pivot.

Balanced moments — nope, not had one of those for a while...

Think of the extra force you need to open a door by pushing it near the hinge compared to at the handle — the distance from the pivot is less, so you need more force to get the same moment. Good times.

Moments, Stability and Pendulums

On the last page we met total clockwise moments being balanced by total anticlockwise moments. Which is all very nice and convenient. But what happens if that isn't the case, I hear you cry. Read on my friend...

If the Moments Acting on an Object aren't Equal the Object will Turn

If the Total Anticlockwise Moments do not equal the Total Clockwise Moments, there will be a Resultant Moment ...so the object will turn.

Low and Wide Objects are Most Stable

Unstable objects tip over easily — stable ones don't. The position of the centre of mass (p.94) is all-important.

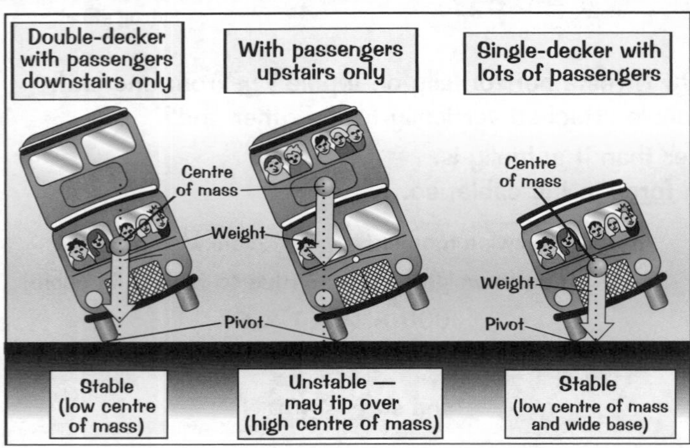

1) The most stable objects have a wide base and a low centre of mass.
2) An object will begin to tip over if its centre of mass moves beyond the edge of its base.
3) Again, it's because of moments — if the line of action of the weight of the object lies outside of the base of the object, it'll cause a resultant moment. This will tip the object over.
4) Lots of objects are specially designed to give them as much stability as possible. For example, a Bunsen burner has a wide, heavy base to give it a low centre of mass — this makes it harder to knock over.

The Time for One Pendulum Swing Depends on its Length

1) A simple pendulum is made by suspending a weight from a piece of string. When you pull back a pendulum and let it go, it will swing back and forth.
2) The time taken for the pendulum to swing from one side to the other and back again is called the time period.
3) The time period for each swing of a given pendulum is the always the same — this is what makes pendulums perfect for keeping time in clocks.
4) The time period can be calculated using this formula:

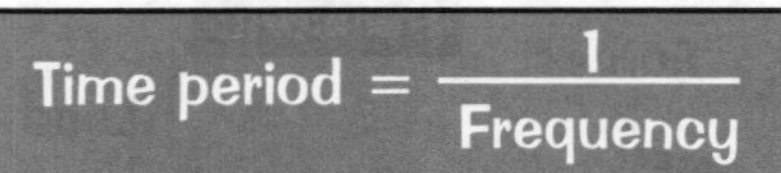

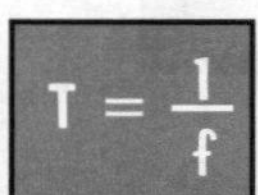

Where:
T = the time period in seconds (s)
f = frequency of the pendulum in hertz (Hz)

5) The time period of a pendulum depends on its length. The longer the pendulum, the greater the time period. So the shorter the length, the shorter the time period.
6) As well as being using in old-style clocks, pendulums have many other (more fun) uses. For example, playground swings are pendulums. Any fairground rides that swing you back and forth are pendulums too. Hooray for pendulums.

You are feeling very sleepy, verrrrry sleeeeepy...

So there you go, the science behind the dangers of the age-old, time-passing activity of 'stool-swinging'. If the centre of mass of you and the stool falls outside of the stool's base, then you're heading for a fall. Ouch.

Circular Motion

If it wasn't for circular motion our little planet would just be wandering aimlessly around the Universe. And as soon as you launched a satellite, it'd just go flying off into space. Hardly ideal.

Circular Motion — Velocity is Constantly Changing

1) Velocity is both the speed and direction of an object (p.40).
2) If an object is travelling in a circle it is constantly changing direction. This means its velocity is constantly changing (but not its speed) — so the object is accelerating (p.40). This acceleration is towards the centre of the circle.
3) There must be a resultant force acting on the object causing this acceleration (p.45). This force acts towards the centre of the circle.
4) This force that keeps something moving in a circle is called a centripetal force.

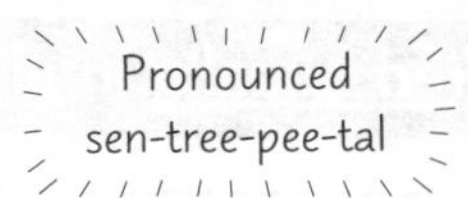

The object's acceleration changes the direction of motion but not the speed.

The force causing the acceleration is always towards the centre of the circle.

In the exam, you could be asked to say which force is actually providing the centripetal force in a given situation. It can be tension, or friction, or even gravity.

A car going round a bend:

1) Imagine the bend is part of a circle — the centripetal force is towards the centre of the circle.
2) The force is from friction between the car's tyres and the road.

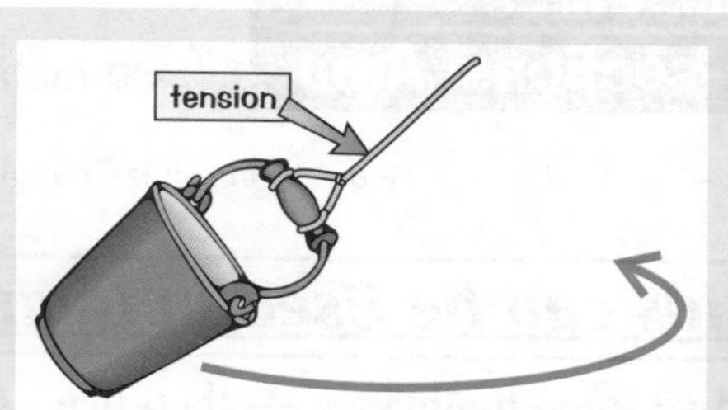

A bucket whirling round on a rope:
The centripetal force comes from tension in the rope. Break the rope, and the bucket flies off at a tangent.

A spinning fairground ride:
The centripetal force comes from tension in the spokes of the ride.

Centripetal Force depends on Mass, Speed and Radius

1) The faster an object's moving, the bigger the centripetal force has to be to keep it moving in a circle.
2) The larger the mass of the object, the bigger the centripetal force has to be to keep it moving in a circle.
3) And you need a larger force to keep something moving in a smaller circle — it has 'more turning' to do.

Example: Two cars are driving at the same speed around the same circular track. One has a mass of 900 kg, the other has a mass of 1200 kg. Which car has the larger centripetal force?

The three things that mean you need a bigger centripetal force are: more speed, more mass, smaller radius of circle.

In this example, the speed and radius of circle are the same — the only difference is the masses of the cars. So you don't need to calculate anything — you can confidently say:

The 1200 kg car (the heavier one) must have the larger centripetal force.

Circular motion — get round to learning it...

To understand this, you need to learn that constant change in direction means constant acceleration. Velocity is a vector — it has direction, and acceleration is change in velocity. When there's acceleration, there's force (see, easy).

Hydraulics

Oh my word, hydraulics. I have to say that word sounds a little scary, but it's not all that bad really. It's all just about how we can use the properties of liquids to our advantage. Mwahaha.

See page 9 for more on liquids.

Liquids are Virtually Incompressible

1) Liquids are virtually incompressible — you can't squash them, their volume and density stay the same.
2) Because liquids are incompressible and can flow, a force applied to one point in the liquid will be transmitted (passed) to other points in the liquid.
3) Imagine a balloon full of water with a few holes in it. If you squeeze the top of the balloon, the water will squirt out of the holes. This shows that force applied to the water at the top of the balloon is transmitted to the water in other parts of the balloon. This also shows that pressure can be transmitted throughout a liquid.

Pressure in a liquid is transmitted equally in all directions.

Pressure and force are linked — see the formula below.

Pressure is the Force per Unit Area

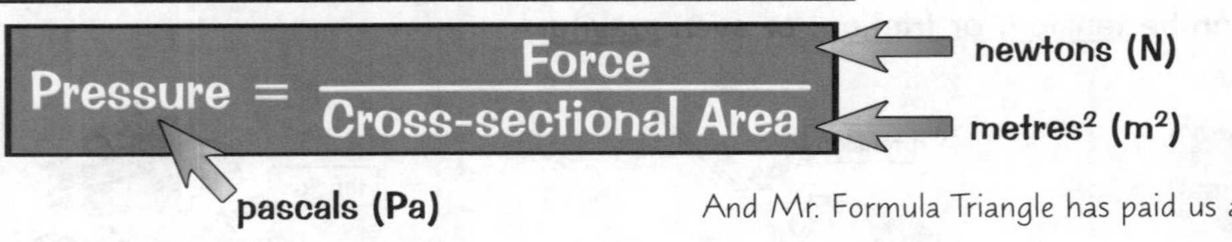

$$\text{Pressure} = \frac{\text{Force}}{\text{Cross-sectional Area}}$$

And Mr. Formula Triangle has paid us a visit...

The Pressure in Liquids can be Used in Hydraulic Systems

1) Hydraulic systems are used as force multipliers — they use a small force to produce a bigger force. They do this using liquid and a sneaky trick with cross-sectional areas.
2) The diagram to the right shows a simple hydraulic system.
3) The system has two pistons, one with a smaller cross-sectional area than the other. Pressure is transmitted equally through a liquid — so the pressure at both pistons is the same.
4) Pressure = force ÷ area, so at the 1st piston, a pressure is exerted on the liquid using a small force over a small area. This pressure is transmitted to the 2nd piston.
5) The 2nd piston has a larger area, and so as force = pressure × area, there will be a larger force.
6) Hydraulic systems are used in all sorts of things, e.g. car braking systems, hydraulic car jacks, manufacturing and deployment of landing gear on some aircraft.

Small force
Piston 1
Piston 2
Small cross-sectional area
Large cross-sectional area
Same pressure
Large force
Liquid

EXAMPLE: To the right is a diagram showing a simple hydraulic system. A force of 15 N is applied to the first piston which has a cross-sectional area of 0.0005 m².

a) Calculate the pressure created on the first piston.

b) Calculate the force acting on the second piston if its cross-sectional area is 0.0012 m².

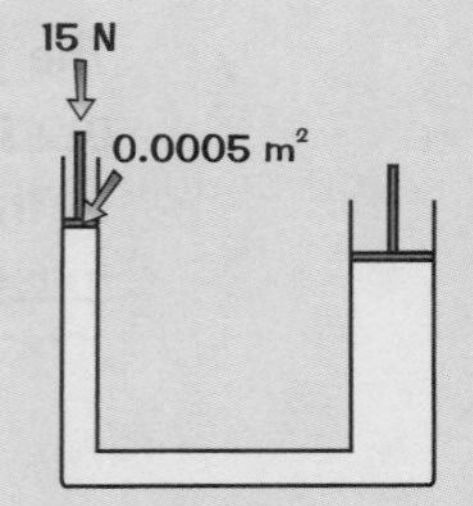

ANSWER: a) $P = F \div A = 15 \div 0.0005 = \underline{30\ 000\ \text{Pa}}$ (or 30 000 N/m²)

b) Pressure at first piston = pressure at second piston, so $F = P \times A = 30\ 000 \times 0.0012 = \underline{36\ \text{N}}$

With all this talk of hydraulics I'm really feeling under pressure...

See, that wasn't too bad was it? Maybe-ish. Hydraulics is all about using a liquid to make a larger force from a smaller one (or in some cases vice versa). It's all thanks to those hard-to-compress liquids, with their pressure-transmitting capabilities. I'll never look at glass of water in the same way again.

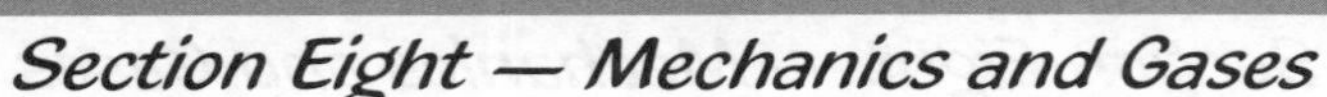

Gas Behaviour

And now for two pages on the behaviour of gas — imaginatively titled 'Gas Behaviour' and 'More on Gas Behaviour'. Funny thing is, it's actually pretty interesting, but don't take my word for it — read it for yourself.

Kinetic Theory Says Gases are Randomly Moving Particles

1) Kinetic theory says that gases consist of very small particles. Which they do — oxygen consists of oxygen molecules, neon consists of neon atoms, etc.
2) These particles are constantly moving in completely random directions.
3) They constantly collide with each other and with the walls of their container. When they collide, they bounce off each other, or off the walls.
4) The particles hardly take up any space. Most of the gas is empty space.

Absolute Zero is as Cold as Stuff Can Get — 0 Kelvin

1) If you increase the temperature of something, you give its particles more kinetic energy — they move about more quickly or vibrate more. In the same way, if you cool a substance down, you're reducing the kinetic energy of the particles.
2) The coldest that anything can ever get is –273 °C — this temperature is known as absolute zero. At absolute zero, atoms have as little kinetic energy as it's possible to get.
3) Absolute zero is the start of the Kelvin scale of temperature.
4) A temperature change of 1 °C is also a change of 1 kelvin. The two scales are pretty similar — the only difference is where the zero is.
5) To convert from degrees Celsius to kelvins, just add 273. And to convert from kelvins to degrees Celsius, just subtract 273.

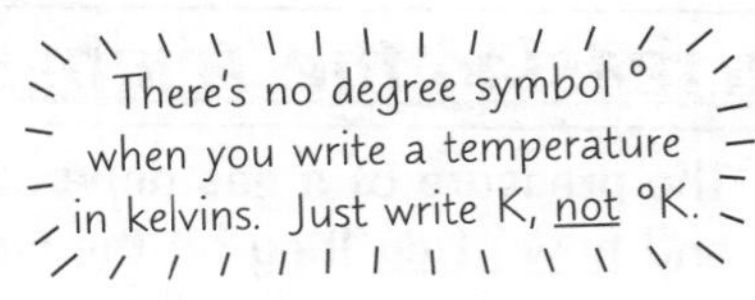

	Absolute zero	Freezing point of water	Boiling point of water
Celsius scale	–273 °C	0 °C	100 °C
Kelvin scale	0 K	273 K	373 K

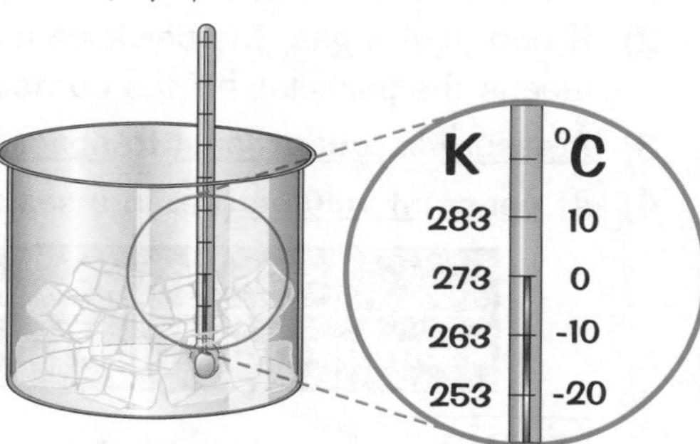

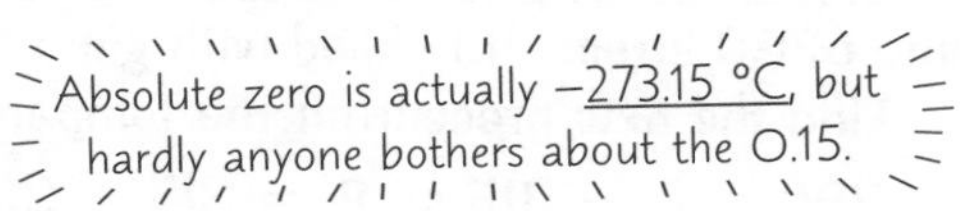

Kinetic Energy is Proportional to Temperature

1) Anything that's moving (e.g. a bunch of particles) has kinetic energy.
2) If you increase the temperature of a gas, you give its particles more energy, so as you heat up a gas the average speed of its particles increases.
3) In fact, if you double the absolute temperature (measured in kelvins), you double the average kinetic energy of the particles:

The temperature of a gas (in kelvins) is proportional to the average kinetic energy of its particles.

Absolute zero — nought, zilch, not a sausage...

Absolute temperature (the Kelvin scale) is handy because it means that temperatures can't be negative. Some equations (like those on the next page) only work using kelvin, so make sure you understand it. Confusingly, the guy who came up with the Kelvin scale was called William Thomson (who became Lord Kelvin).

More on Gas Behaviour

A Decrease in Volume Gives an Increase in Pressure

1) As gas particles move about, they bang into each other and whatever else happens to get in the way.
2) Gas particles have some mass, so when they collide with something, they exert a force on it. In a sealed container, gas particles smash against the container's walls — creating an outward pressure.
3) If you put the same amount of gas in a bigger container, the pressure will decrease, cos there'll be fewer collisions between the gas particles and the container's walls. When the volume's reduced, the particles get more squashed up and so they hit the walls more often, hence the pressure increases.
4) So the volume of a gas is inversely proportional to its pressure at a constant temperature — e.g. if you halve the volume, you double the pressure.
5) At constant temperature: **pressure × volume = constant**

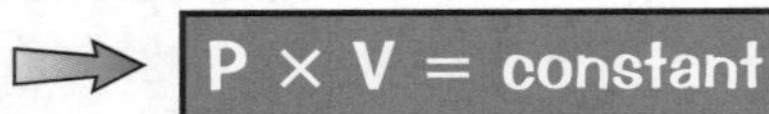

You can also write the equation as $P_1 \times V_1 = P_2 \times V_2$ (where P_1 and V_1 are your starting conditions and P_2 and V_2 are your final conditions). Writing it like that is much more useful a lot of the time.

> **EXAMPLE:** A gas at a constant temperature in a 50 ml container has a pressure of 1.2 atm. Find the new pressure if the container volume is reduced to 40 ml.
>
> **ANSWER:** $P_1 \times V_1 = P_2 \times V_2$ gives: $1.2 \times 50 = P_2 \times 40$ so $P_2 = 60 \div 40 =$ 1.5 atm

atm = atmosphere, a unit of pressure.

Increasing the Temperature Increases the Pressure

1) The pressure of a gas depends on how fast the particles are moving and how often they hit the walls of the container they're in.
2) If you heat a gas, the particles move faster and have more kinetic energy. This increase in kinetic energy means the particles hit the container walls harder and more often, creating more pressure.
3) Pressure is proportional to absolute temperature — doubling the temperature (in K), doubles the pressure.
4) At constant volume (i.e. in a sealed container):

$$\frac{\text{pressure}}{\text{temperature (in K)}} = \text{constant}$$

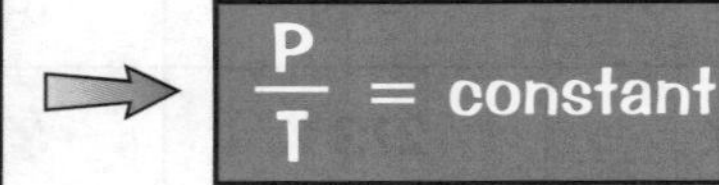

You can also write the equation as: $P_1/T_1 = P_2/T_2$.

> **EXAMPLE:** A container has a volume of 30 litres. It is filled with gas at a pressure of 1 bar and a temperature of 290 K. Find the new pressure if the temperature is increased to 315 K.
>
> **ANSWER:** $P_1/T_1 = P_2/T_2$ gives: $1 \div 290 = P_2 \div 315$ so $P_2 = 315 \div 290 =$ 1.09 bar

1 bar is roughly the same as 1 atm.

Increasing the Temperature Increases the Volume

1) If a gas stays at a constant pressure, then heating it up increases its volume — the molecules are further apart so collisions happen less frequently, but with more force (because they have more kinetic energy).
2) Volume is proportional to absolute temperature. Doubling the temperature (in K), doubles the volume.
3) At constant pressure:

$$\frac{\text{volume}}{\text{temperature (in K)}} = \text{constant}$$

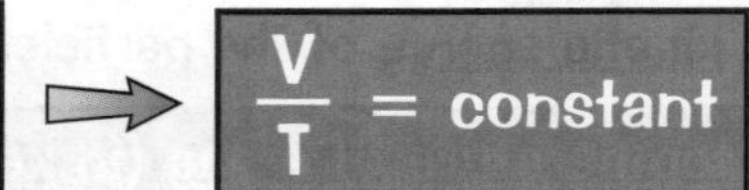

You can also write the equation as: $V_1/T_1 = V_2/T_2$.

> **EXAMPLE:** A gas at constant pressure, with a temperature of 270 K has a volume of 24 litres. Find the new volume if the temperature is increased to 315 K.
>
> **ANSWER:** $V_1/T_1 = V_2/T_2$ gives: $24 \div 270 = V_2 \div 315$ so $V_2 = (24 \div 270) \times 315 =$ 28 litres

Less space, more collisions, more pressure — just like London...

These equations apply to so-called ideal gases. Ideal gases are those that are 'well behaved'. Nice gases. Stay.

Revision Summary for Section Eight

Well, here we are again. It's time for another round of questions. You've probably had enough of me wittering on by now — so I'll leave you to get stuck in. Good luck, chaps.

1) Sarah is levering the lid off a can of paint using a screwdriver. She places the tip of the 20 cm long screwdriver under the can's lid and applies a force of 10 N on the end of the screwdriver's handle. Suggest two ways that Sarah could increase the moment about the pivot point (the side of the can).

2)* The diagram on the right shows a seesaw. Calculate the moment exerted if a force of 440 N is applied at the seat at a distance of 1.75 m from the pivot.

440 N

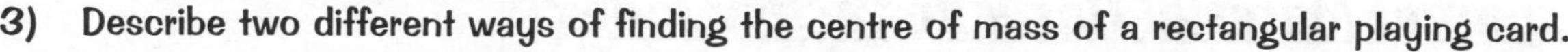

3) Describe two different ways of finding the centre of mass of a rectangular playing card.

4)* Arthur weighs 600 N and is sitting on a seesaw 1.5 m from the pivot point. His friend Caroline weighs 450 N and sits on the seesaw so that it balances. How far from the pivot point is Caroline sitting?

5) Explain why levers can be called force multipliers.

6) Give three situations where you use a simple lever.

7) Give two features of a Bunsen burner that make it difficult to tip over.

8)* Calculate the time period of a pendulum swinging with a frequency of 10 Hz.

9) Explain how you could change a pendulum to increase its time period.

10) A cyclist is moving at a constant speed of 5 m/s around a circular track.
   a) Is the cyclist accelerating? Explain your answer.
   b) What force keeps the cyclist travelling in a circle? Where does this force come from?
   c) What will happen to the size of this force if the same cyclist travels at a constant speed of 5 m/s around a different circular track that has a larger radius?

11)* A force of 20 N is applied to a piston in a hydraulic system with a cross-sectional area 0.25 m^2. Calculate the pressure applied to the piston.

12)* A hydraulic system is shown in the diagram on the right. What is the force acting on piston B if it has a cross sectional area of 0.36 m^2?

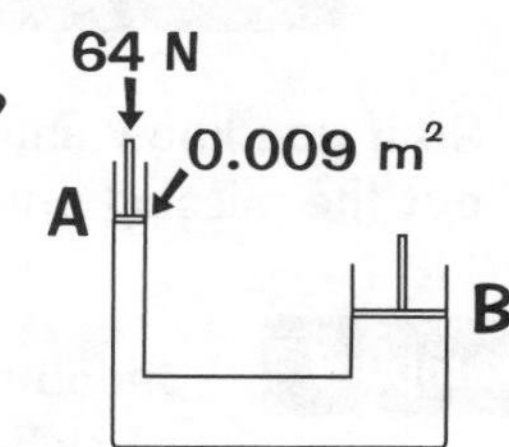

13) Give one use of a hydraulic system.

14) What's absolute zero in °C? What does absolute zero mean in terms of kinetic energy of particles?

15)* Calculate the following temperature conversions:
   a) –89 °C into K b) 120 °C into K c) 5 K into °C d) 312 K into °C

16) What creates the pressure that a gas exerts on the walls of its container?

17)* A gas syringe holds 160 cm^3 of oxygen at a pressure of 101 325 Pa. The system is kept at a constant temperature of 300 K. Calculate the pressure of the gas when compressed to a volume of 40 cm^3.

18) Give the equation for the relationship between the temperature and pressure of a gas in a sealed rigid container.

19)* A container of fixed volume 0.0005 m^3 is filled with gas at a pressure of 5.0 atm. The container is cooled from 25 °C to -100 °C. What will be the new pressure inside the container?

20) What two ways are there to increase the pressure of a gas?

* Answers on page 140.

Refractive Index

You may have noticed that transparent materials can do some funny things to light. It might be to do with witchcraft, but then again it might also be to do with refractive index...

Refraction is Caused by the Waves Changing Speed

1) Refraction is when waves change direction as they enter a different medium (see page 29). This is caused by the change in density from one medium to the other — which changes the speed of the waves.
2) When light enters glass or plastic it slows down — to about 2/3 of its speed in air.

Every Transparent Material Has a Refractive Index

1) The refractive index of a medium is the ratio of speed of light in a vacuum to speed of light in that medium.

$$\text{refractive index } (n) = \frac{\text{speed of light in a vacuum } (c)}{\text{speed of light in the medium } (v)} \qquad n = \frac{c}{v} \qquad (c = 3 \times 10^8 \text{ m/s})$$

2) Light slows down a lot in glass, so the refractive index of glass is high (around 1.5). The refractive index of water is lower (around 1.33) — so light slows down less in water than in glass.
3) The speed of light in air is about the same as in a vacuum, so the refractive index of air is 1.
4) When an incident ray passes from air into another material, the angle of refraction of the ray depends on the refractive index of the material.
5) The angle of incidence, i, angle of refraction, r, and refractive index, n, are all linked.

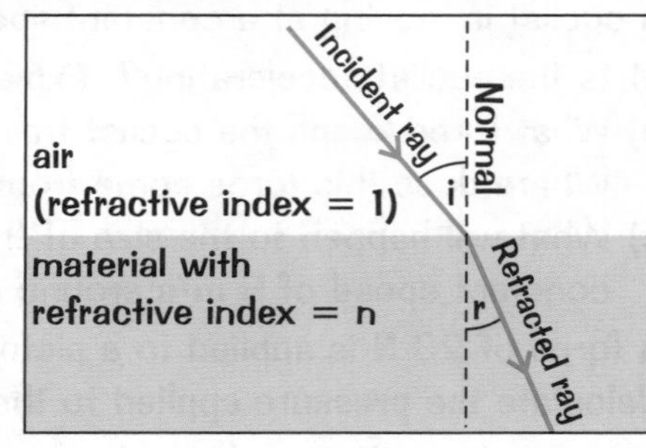

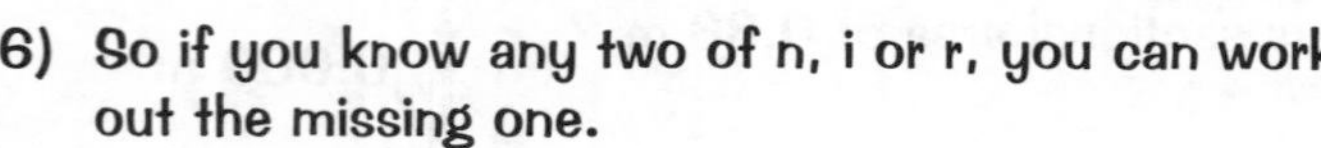

You might see this equation called Snell's Law.

6) So if you know any two of n, i or r, you can work out the missing one.

Example 1

Jacob does an experiment to find out the refractive index of his strawberry flavour jelly. He finds that when the angle of incidence for a light beam travelling into his jelly is 42°, the angle of refraction is 35°. What is the refractive index of this particular type of jelly?

$$n = \frac{\sin i}{\sin r} \qquad \sin i = \sin 42 = 0.67 \qquad \sin r = \sin 35 = 0.57 \quad \Rightarrow \quad \text{so } n_{jelly} = \frac{0.67}{0.57} = \underline{1.18}$$

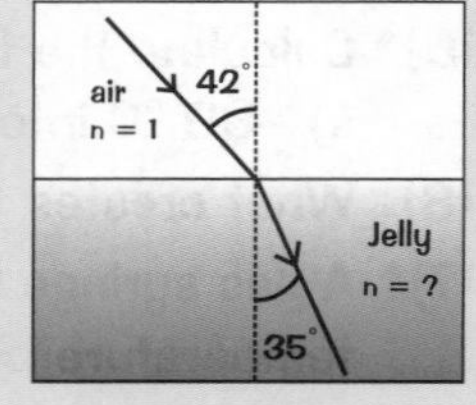

Example 2

A beam of light travels from air into water (refractive index n = 1.33). The angle of incidence is 23°. Calculate the angle of refraction to the nearest degree.

$$\sin r = \frac{\sin i}{n} = \frac{\sin 23}{1.33} = 0.29 \quad \Rightarrow \quad \text{so } r = \sin^{-1}(0.29) = \underline{17°}$$

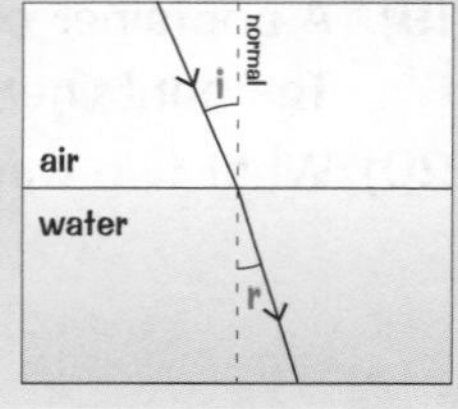

All that glitters has a high refractive index...

It's really important that you remember that the angles of incidence and refraction are always measured from the normal to the light ray, not from the boundary between materials to the light ray. Got it? Good.

Prisms and Total Internal Reflection

Now you've got your head well and truly around refractive index (see previous page), you can see some of the stuff it can explain (like dispersion) and the things we can use it for (like optical fibres).

Refractive Index Explains Dispersion

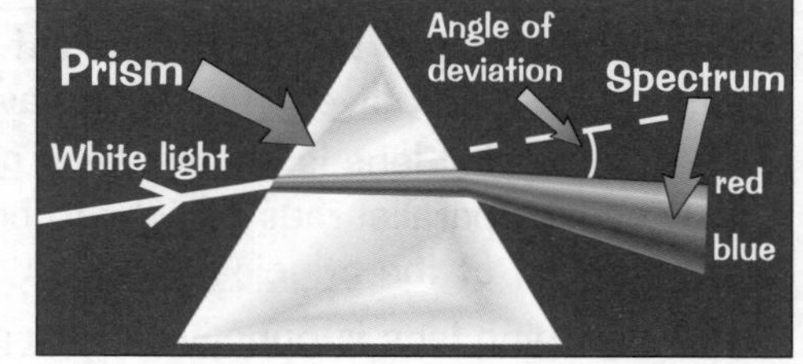

1) Different colours of light are refracted by different amounts.
2) This is because they travel at slightly different speeds in any given medium (but the same speed in a vacuum).
3) Any material has a different refractive index for each different speed (colour) of light.
4) Red light slows down the least when it travels from air into glass, so it is refracted the least and has the lowest refractive index (1.514). Blue light has a higher refractive index (1.523) so is refracted more.
5) A prism can be used to make the different colours of white light emerge at different angles.
6) This produces a spectrum showing all the colours of the rainbow. This effect is called **DISPERSION**.

Light can be Sent Along Optical Fibres Using Total Internal Reflection

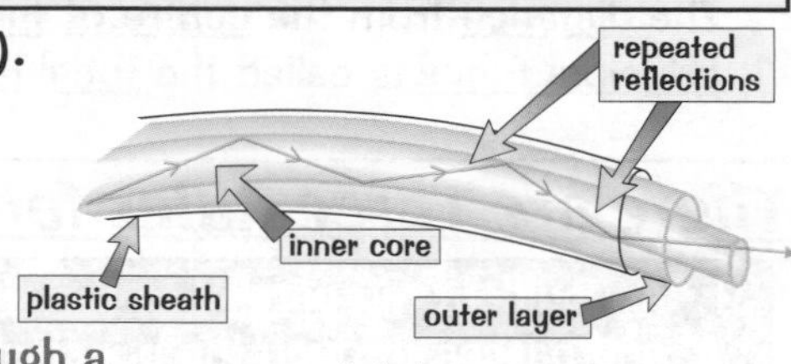

1) Optical fibres can carry visible light over long distances (see p.34).
2) They work by bouncing waves off the sides of a thin inner core of glass or plastic. The wave enters one end of the fibre and is reflected repeatedly until it emerges at the other end.
3) Optical fibres work because of total internal reflection.
4) Total internal reflection can only happen when a wave travels through a dense substance like glass or water towards a less dense substance like air.
5) It all depends on whether the angle of incidence is bigger than the critical angle...

Remember, angle of reflection, r, equals the angle of incidence, i.

If the angle of incidence (i) is...

The angle of incidence (i) and the angle of reflection (r) are always measured from the normal (a line at right angles to the surface).

i r

Critical angle i — slightly stronger reflected ray

i — total internal reflection

...LESS than Critical Angle:- Most of the light passes out but a little bit of it is internally reflected.

...EQUAL to Critical Angle:- The emerging ray comes out along the surface. There's quite a bit of internal reflection.

...GREATER than Critical Angle:- No light comes out. It's all internally reflected, i.e. total internal reflection.

The Value of the Critical Angle Depends on the Refractive Index

1) A dense material with a high refractive index has a low critical angle.
2) If a material has a high refractive index, it will totally internally reflect more light — more light will be incident at an angle bigger than the critical angle.
3) You can find the critical angle, C, using this equation:

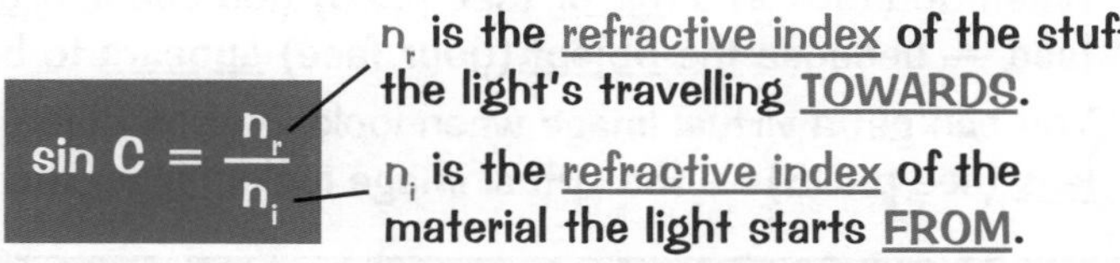

$$\sin C = \frac{n_r}{n_i}$$

n_r is the refractive index of the stuff the light's travelling TOWARDS.

n_i is the refractive index of the material the light starts FROM.

4) Quite a lot of the time you'll be looking at light travelling from some material into air. The refractive index of air is about 1 (see previous page) so the equation above becomes: $\sin C = 1 \div n$.

Lenses and Images

This bit is about how light acts when it hits a lens. Be ready for lots of diagrams on the next few pages.

Different Lenses Produce Different Kinds of Image

Lenses form images by refracting light and changing its direction. There are two main types of lens — converging and diverging. They have different shapes and have opposite effects on light rays.

1) A converging lens is convex — it bulges outwards. It causes parallel rays of light to converge (move together) at the principal focus.
2) A diverging lens is concave — it caves inwards. It causes parallel rays of light to diverge (spread out).
3) The axis of a lens is a line passing through the middle of the lens.
4) The principal focus of a converging lens is where rays hitting the lens parallel to the axis all meet.
5) The principal focus of a diverging lens is the point where rays hitting the lens parallel to the axis appear to all come from — you can trace them back until they all appear to meet up at a point behind the lens.
6) There is a principal focus on each side of the lens. The distance from the centre of the lens to the principal focus is called the focal length.

converging lens
principal focus
axis
diverging lens
axis
virtual principal focus
virtual ray

The principal focus is also known as the focal point.

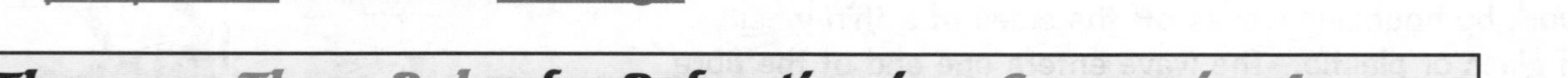

There are Three Rules for Refraction in a Converging Lens...

1) An incident ray parallel to the axis refracts through the lens and passes through the principal focus on the other side.
2) An incident ray passing through the principal focus refracts through the lens and travels parallel to the axis.
3) An incident ray passing through the centre of the lens carries on in the same direction.

See next page for more on this.

... And Three Rules for Refraction in a Diverging Lens

1) An incident ray parallel to the axis refracts through the lens, and travels in line with the principal focus (so it appears to have come from the principal focus).
2) An incident ray passing through the lens towards the principal focus refracts through the lens and travels parallel to the axis.
3) An incident ray passing through the centre of the lens carries on in the same direction.

See next page for more on this.

The neat thing about these rules is that they allow you to draw ray diagrams without bending the rays as they go into the lens and as they leave the lens. You can draw the diagrams as if each ray only changes direction once, in the middle of the lens (see next page).

Lenses can Produce Real and Virtual Images

1) A real image is where the light from an object comes together to form an image on a 'screen' — like the image formed on an eye's retina (the 'screen' at the back of an eye).
2) A virtual image is when the rays are diverging, so the light from the object appears to be coming from a completely different place.

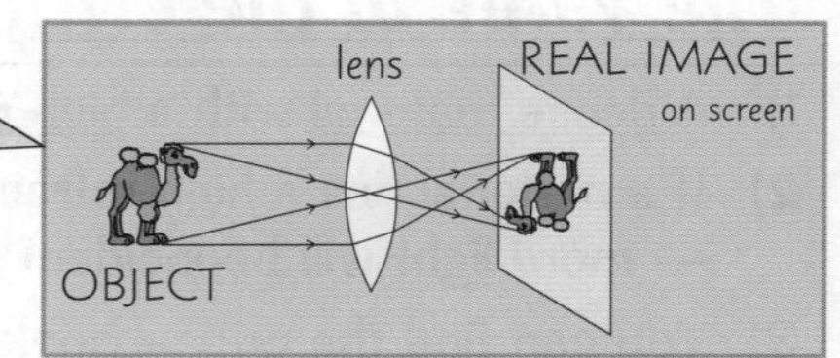

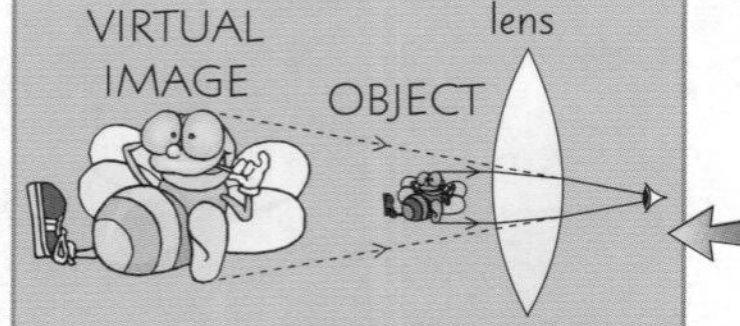

3) When you look in a mirror (see p.28) you see a virtual image of your face — because the object (your face) appears to be behind the mirror.
4) You can get a virtual image when looking at an object through a magnifying lens (see p.106) — the virtual image looks bigger than the object actually is.

To describe an image properly, you need to say 3 things: 1) How big it is compared to the object; 2) Whether it's upright or inverted (upside down) relative to the object; 3) Whether it's real or virtual.

Lenses

You might have to draw a ray diagram of refraction through a lens. Follow the instructions very carefully...

Draw a Ray Diagram for an Image Through a Converging Lens

In ray diagrams, this represents a convex lens...

1) Pick a point on the top of the object. Draw a ray going from the object to the lens parallel to the axis of the lens.
2) Draw another ray from the top of the object going right through the middle of the lens.
3) The incident ray that's parallel to the axis is refracted through the principal focus (F). Draw a refracted ray passing through the principal focus.
4) The ray passing through the middle of the lens doesn't bend.
5) Mark where the rays meet. That's the top of the image.
6) Repeat the process for a point on the bottom of the object. When the bottom of the object is on the axis, the bottom of the image is also on the axis.

If you really want to draw a third incident ray passing through the principal focus on the way to the lens, you can (refract it so that it goes parallel to the axis). In the exam, you can get away with two rays, so no need to bother with three.

Distance from the Lens Affects the Image

1) An object at 2F will produce a real, inverted (upside down) image the same size as the object, and at 2F.

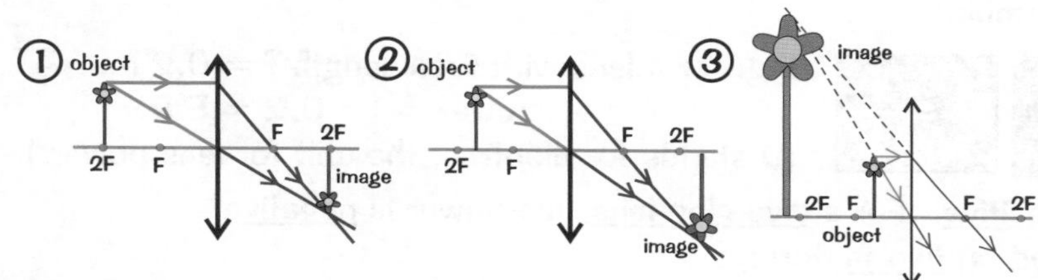

2) Between F and 2F it'll make a real, inverted image bigger than the object, and beyond 2F.
3) An object nearer than F will make a virtual image the right way up, bigger than the object, on the same side of the lens.

Draw a Ray Diagram for an Image Through a Diverging Lens

...and this represents a concave lens.

1) Pick a point on the top of the object. Draw a ray going from the object to the lens parallel to the axis of the lens.
2) Draw another ray from the top of the object going right through the middle of the lens.
3) The incident ray that's parallel to the axis is refracted so it appears to have come from the principal focus. Draw a ray from the principal focus. Make it dotted before it reaches the lens.
4) The ray passing through the middle of the lens doesn't bend.
5) Mark where the refracted rays meet. That's the top of the image.
6) Repeat the process for a point on the bottom of the object. When the bottom of the object is on the axis, the bottom of the image is also on the axis.

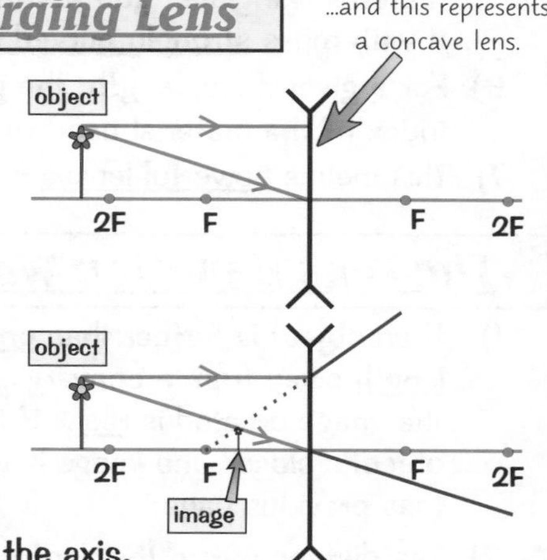

Again, if you really want to draw a third incident ray in the direction of the principal focus on the far side of the lens, you can. Remember to refract it so that it goes parallel to the axis. In the exam, you can get away with two rays. Choose whichever two are easiest to draw — don't try to draw a ray that won't actually pass through the lens.

The Image is Always Virtual

A diverging lens always produces a virtual image. The image is the right way up, smaller than the object and on the same side of the lens as the object — no matter where the object is.

Warning — too much revision can cause attention to diverge...

Get busy practising drawing those ray diagrams. Like riding a bike, you learn by doing — so jump on it.

Magnification and Power

Magnifying Glasses Use Converging Lenses

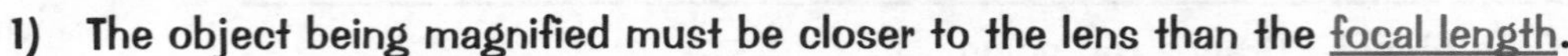

Magnifying glasses work by creating a magnified virtual image (see p.104).

1) The object being magnified must be closer to the lens than the focal length.
2) Since the image produced is a virtual image, the light rays don't actually come from the place where the image appears to be.
3) Remember "you can't project a virtual image onto a screen" — a useful phrase to use in the exam if they ask you about virtual images.

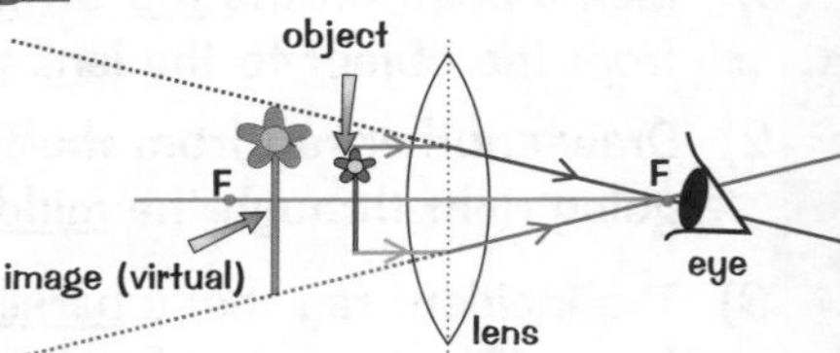

Learn the Magnification Formula

You can use the magnification formula to work out the magnification produced by a lens at a given distance:

$$\text{Magnification} = \frac{\text{image height}}{\text{object height}}$$

Example: A coin with diameter 14 mm is placed a certain distance behind a magnifying lens. The virtual image produced has a diameter of 35 mm. What is the magnification of the lens at this distance?

magnification = 35 ÷ 14
= 2.5

A Powerful Lens has a Short Focal Length

1) Focal length is related to the power of the lens. The more powerful the lens, the more strongly it converges rays of light, so the shorter the focal length (see p.104).
2) The power of a lens is given by the formula:

$$\text{Power (D)} = \frac{1}{\text{Focal length (m)}}$$

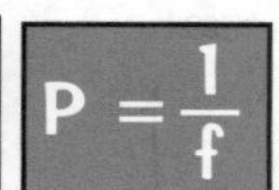

E.g. for a lens with focal length f = 0.2 m,
power = 1 ÷ 0.2 = 5 D.
(D stands for dioptres, the unit for lens power.)

3) For a converging lens, the power is positive. For a diverging lens, the power is negative.
4) The focal length of a lens is determined by two factors:
   a) the refractive index of the lens material,
   b) the curvature of the two surfaces of the lens.
5) To make a more powerful lens from a certain material like glass, you just have to make it with more strongly curved surfaces.
6) For a given focal length, the greater the refractive index of the material used to make the lens, the flatter the lens will be.
7) This means powerful lenses can be made thinner by using materials with high refractive indexes (see p.102).

The Lens Equation Works for Converging Lenses

1) If an object is further than one focal length away from a converging lens, the image created is real. If the object's closer, the image is virtual (see previous page).

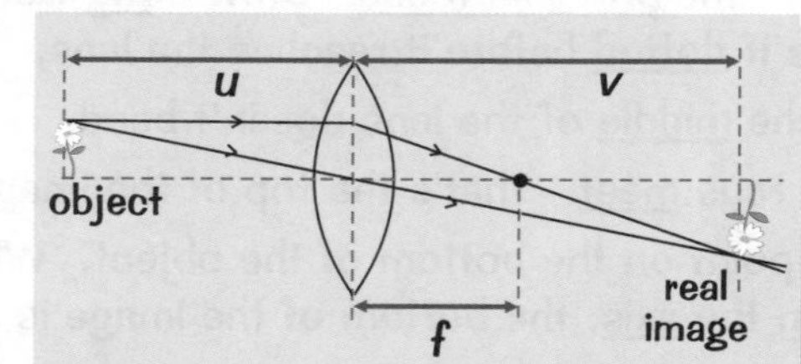

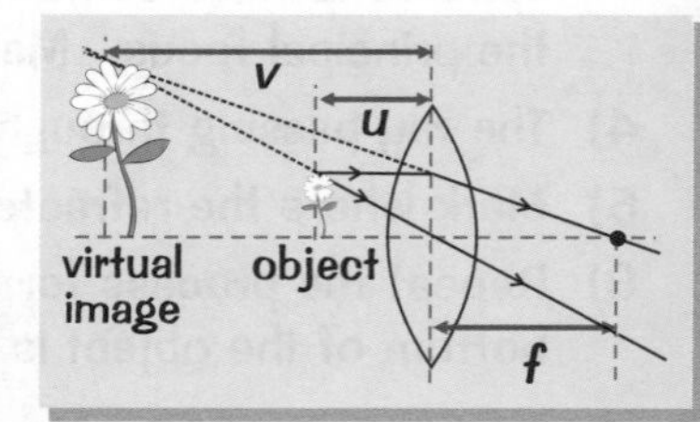

2) The distance from the centre of the lens to the object, u, is the object distance. v is the image distance — the distance between the lens and the image, and f is the focal length. All units are metres (m).
3) The relationship between the position of the object, the position of the image and the focal length relative to the lens is described by the lens equation.

$$\frac{1}{f} = \frac{1}{u} + \frac{1}{v}$$

4) v is positive if the image is real, and negative if the image is virtual.

He's magnificent, that pug...

People with bad eyesight used to have really thick glasses back in the day. Thankfully, nowadays you can get thin high-index plastic lenses that not only look better, but are also much more comfortable to wear. Huzzah.

The Eye

The eye is an absolute marvel of evolution — the way all the different parts work together to form an image and transport it to your brain is quite astonishing... Well, I like it anyway.

You Need to Know the Basic Structure of the Eye

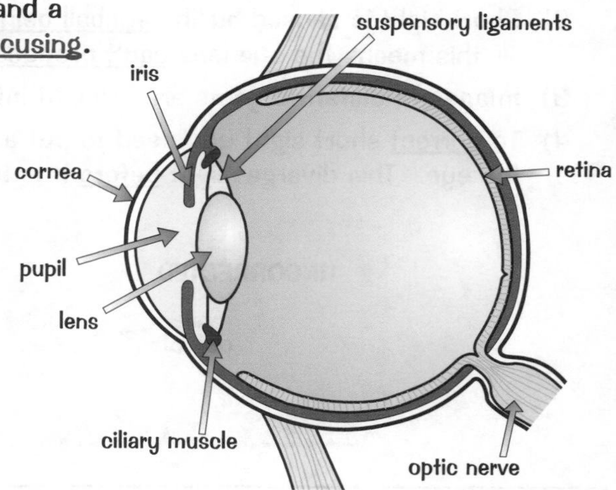

1) The cornea is a transparent 'window' with a convex shape, and a high refractive index. The cornea does most of the eye's focusing.
2) The iris is the coloured part of the eye. It's made up of muscles that control the size of the pupil — the hole in the middle of the iris. This controls the intensity of light entering the eye.
3) The lens changes shape to focus light from objects at varying distances. It's connected to the ciliary muscles by the suspensory ligaments and when the ciliary muscles contract, tension is released and the lens takes on a fat, more spherical shape. When they relax, the suspensory ligaments pull the lens into a thinner, flatter shape.
4) Images are formed on the retina, which is covered in light-sensitive cells. These cells detect light and send signals to the brain via the optic nerve to be interpreted.

The Eye can Focus on Objects Between the Near and Far Points

1) The far point is the furthest distance that the eye can focus comfortably. For normally-sighted people, that's infinity.
2) The near point is the closest distance that the eye can focus on. For adults, the near point is approximately 25 cm.
3) As the eye focuses on closer objects, its power increases — the lens changes shape and the focal length decreases. But the distance between the lens and the image stays the same.

A Camera Forms Images in a Similar Way to the Eye

When you take a photograph of a flower, light from the object (flower) travels to the camera and is refracted by the lens, forming an image on the film.

1) The image on the film is a real image because light rays actually meet there.
2) The image is smaller than the object, because the object's a lot further away than the focal length of the lens.

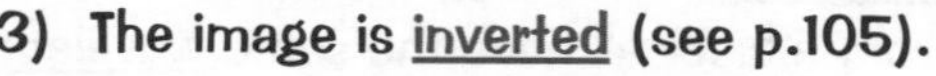

3) The image is inverted (see p.105).

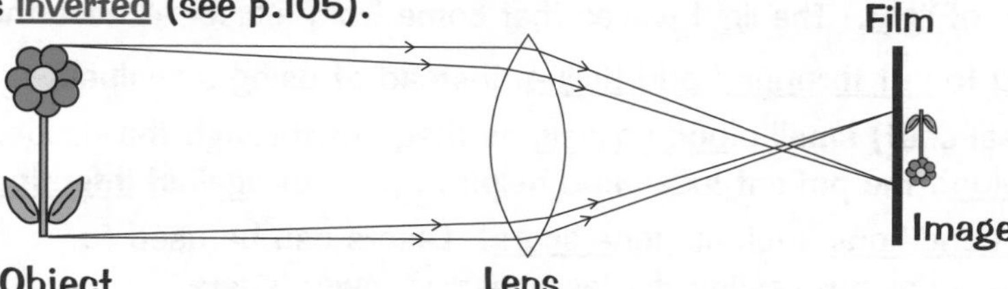

4) The same thing happens in our eye — a real, inverted image forms on the retina. Our very clever brains flip the image so that we see it right way up.
5) The film in a camera, or the CCD in a digital camera (see p.88), are the equivalent of the retina in the eye — they all detect the light focused on them and record it.

Eyes eyes baby...

The light-sensitive cells in the retina at the back of the eye send signals to the brain depending on the amount and colour of the light they've been exposed to. Then your grey matter works out all the different signals, forms an image, puts the image the right way up and figures out what it is. See, I told you it was nifty.

Correcting Vision

Sometimes things go a little awry in the eye department — and that's when physics steps in to save the day.

Short Sight is Corrected with Diverging Lenses

1) Short-sighted people can't focus on distant objects — their far point is closer than infinity (see p.107).
2) Short sight is caused by the eyeball being too long, or by the cornea and lens system being too powerful — this means the eye lens can't produce a focused image on the retina where it is supposed to.
3) Images of distant objects are brought into focus in front of the retina instead.
4) To correct short sight you need to put a diverging lens (with a negative power, see p.106) in front of the eye. This diverges light before it enters the eye, which means the lens can focus it on the retina.

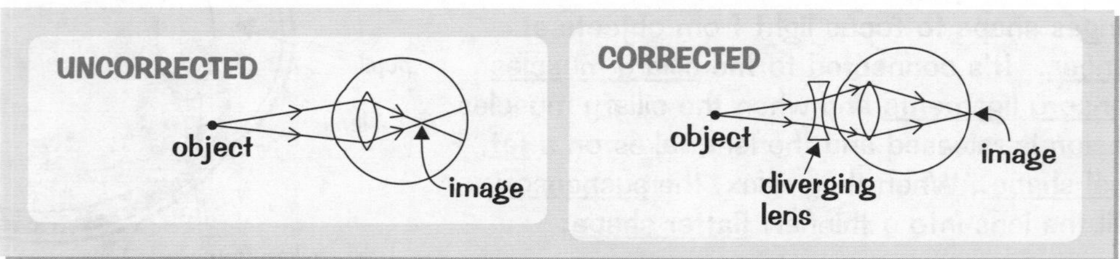

Long Sight is Corrected with Converging Lenses

1) Long-sighted people can't focus clearly on near objects — their near point is further away than normal (25 cm or more, see p.107).
2) Long sight happens when the cornea and lens are too weak or the eyeball is too short.
3) This means that images of near objects are brought into focus behind the retina.
4) To correct long sight a converging lens (with a positive power, see p.106) can be put in front of the eye. The light is refracted and starts to converge before it enters the eye, and the image can be focused on the retina where it belongs.

Many young children are long-sighted — their lenses grow quicker than their eyeballs.

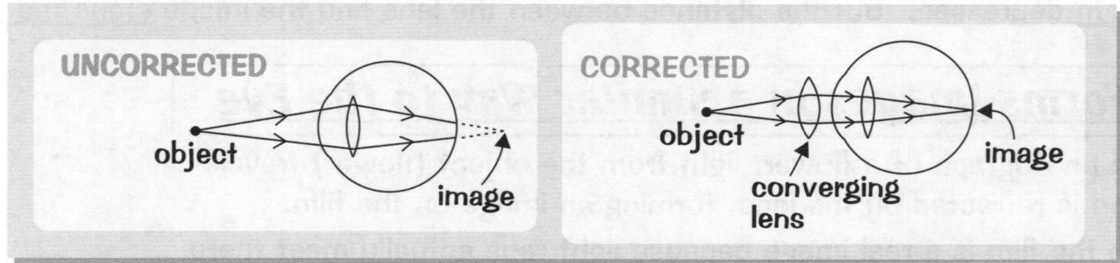

Lasers are Used to Surgically Correct Eye Problems

A laser is an narrow, intense beam of light. The light waves that come from a laser all have the same wavelength.

1) Lasers can be used in surgery to cut through body tissue, instead of using a scalpel.
2) Lasers cauterise (burn and seal shut) small blood vessels as they cut through the tissue. This reduces the amount of blood the patient loses and helps to protect against infection.
3) Lasers are used to treat skin conditions such as acne scars. Lasers can be used to burn off the top layers of scarred skin revealing the less-scarred lower layers.
4) One of the most common types of laser surgery is eye surgery. A laser can be used to vaporise some of the cornea to change its shape — which changes its focusing ability. This can increase or decrease the power of the cornea so that the eye can focus images properly on the retina.

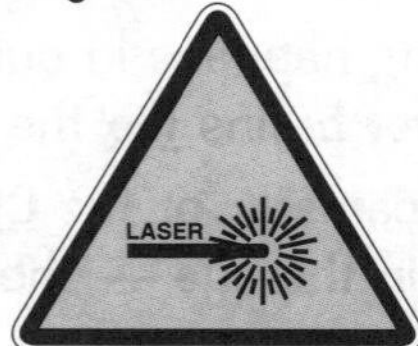

Wear glasses — they give you specs appeal...

The light from a laser can be let out in pulses for even more control during delicate eye surgery. The surgeon can precisely control how much tissue the laser takes off by using the pulses of light to do only a little bit at a time.

Telescopes

Telescopes use converging lenses or curved mirrors to magnify light from space. Here's how.

A Simple Refracting Telescope Uses Two Converging Lenses

1) A simple refracting optical telescope is made up of two convex lenses with different powers — an objective lens and a more powerful eye lens (or eyepiece). The objective lens collects the light from the object being observed and forms an image of it, and the eyepiece magnifies this image so we can view it.
2) The lenses are aligned to have the same principal axis and are placed so that their principal focuses (see p.104) are in the same place.

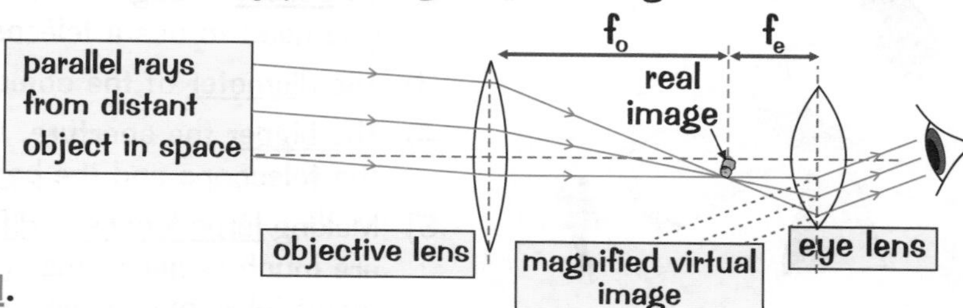

3) Many objects in space are so far away that by the time their light arrives on Earth, the light rays are effectively parallel.
4) The objective lens converges these parallel rays to form a real image between the two lenses.
5) The eyepiece lens is much more powerful than the objective lens (it's much more curved — see p.106). It acts as a magnifying glass on the real image and makes a virtual image — where the light entering the eye lens appears to have come from.
6) The angular magnification, M, of the telescope can be calculated from the focal lengths of the objective lens, f_o, and the eye lens, f_e.

$$\text{Magnification} = \frac{\text{Focal length of objective lens } (f_O)}{\text{Focal length of eye lens } (f_e)}$$

EXAMPLE 1: An astronomer uses a refracting telescope to look at a distant star. If $f_o = 4.5$ m and $f_e = 0.1$ m, find the angular magnification of the telescope.

ANSWER: Angular magnification = $f_o \div f_e = 4.5 \div 0.1 = 45$

EXAMPLE 2: Geoff makes a telescope which has a 0.2 D objective lens and an 8 D eye lens. What is the magnification of Geoff's telescope?

ANSWER: $P = 1 \div f$ (see p.106), so $f = 1 \div P$. So $f_o = 1 \div 0.2 = 5$ m, $f_e = 1 \div 8 = 0.125$ m.
Angular magnification = $f_o \div f_e = 5 \div 0.125 = 40$

See right at the bottom of the page for a shortcut for doing this one.

Most Astronomical Telescopes Use a Concave Mirror

1) Most astronomical telescopes use a concave mirror instead of a convex objective lens.
2) Concave mirrors are shiny on the inside of the curve. Parallel rays of light shining on a concave mirror reflect and converge.
3) Concave mirrors are like a portion of a sphere. The centre of the sphere is the centre of curvature, C.
4) The centre of the mirror's surface is called the vertex.
5) Halfway between the centre of curvature and the vertex is the principal focus, F. These points all lie on the axis.
6) Rays parallel to the mirror's axis, e.g. those from a distant star, reflect and meet at the principal focus (as with lenses).

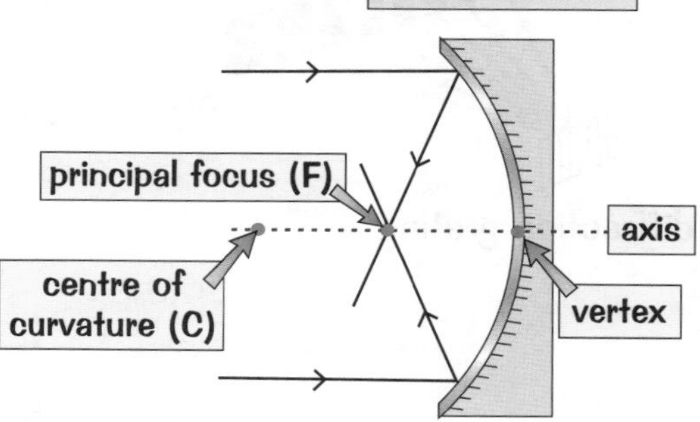

7) By putting a lens near the principal focus of the mirror to act as an eyepiece, you can form a magnified image — just as in the simple refracting telescope above.

Important stuff this — come on, focus, focus...

There's an alternative magnification formula you can use if you're given the powers rather than focal lengths. It'll save faffing about: Magnification = $\text{Power}_{\text{eye}} \div \text{Power}_{\text{objective}}$. (It's just upside down.)

Diffraction and Telescopes

You've met diffraction already, way back on page 29, but this page goes beyond the basics — it's all about the way diffraction affects what we can see with telescopes.

To Detect Faint Sources You Need a Wide Aperture

1) Some objects in the sky are so distant and faint, only a tiny amount of radiation from them reaches us.

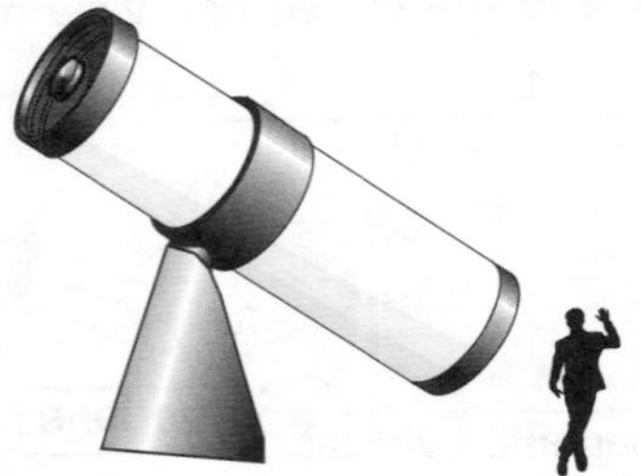

2) To collect enough of the radiation from these objects to see them, you need to use a telescope with a huge objective lens (or mirror).
3) The diameter of the objective lens is called the aperture.
4) The bigger the aperture, the more radiation can get into the telescope and the better the image formed.
5) Making large lenses is difficult and expensive, whereas big mirrors are much easier to make accurately. This is one of the reasons why many telescopes have a concave mirror instead of a lens (see p.109).

Aperture Size Must be Much Larger Than Wavelength

1) Because all waves can diffract when they pass through a gap, radiation entering a telescope spreads out at the edges of the aperture — causing the image to blur.
2) If you looked at radiation from a point source, like a star, when there was no diffraction, you'd see a bright dot.
3) If you used too small an aperture to look at the star, the dot would be dimmer and surrounded by rings which get dimmer the further away from the image they are.

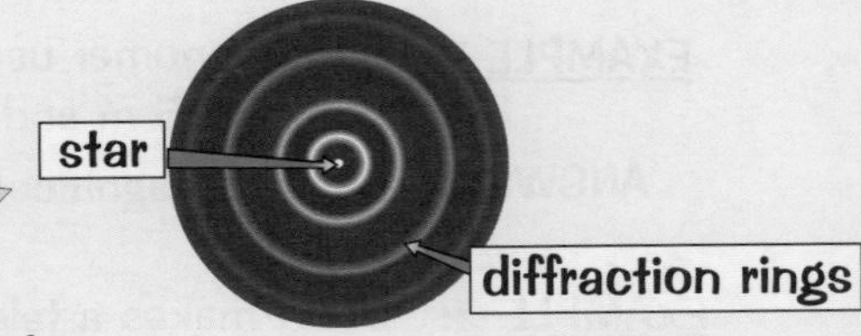

4) The only way round this problem is to have an aperture that's much wider than the wavelength of radiation you want to look at. This way the radiation passes through the aperture and into your telescope with very little diffraction and you get a sharp image.

A Diffraction Grating Can Be Used To Make a Spectrum

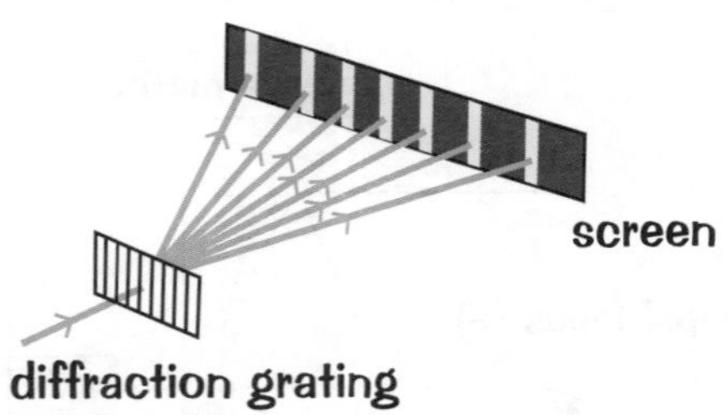

1) A diffraction grating has very narrow slits — small enough to diffract light.
2) When white light passes through the gaps in a diffraction grating, the different wavelengths of coloured light are all diffracted but by different amounts.
3) This creates a spectrum of different coloured light, as shown below.

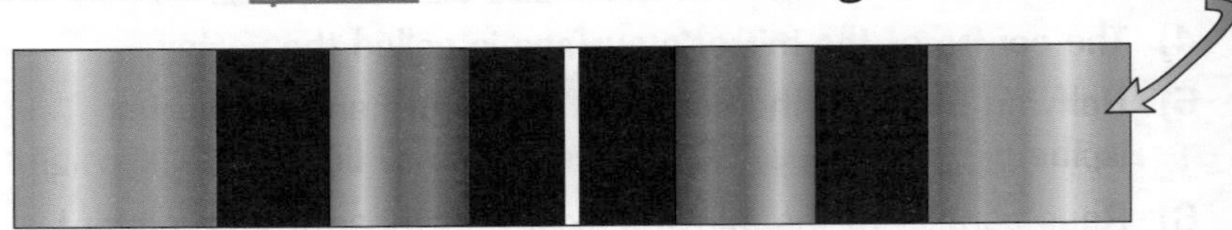

4) Astronomers can use these spectra to analyse the light coming from stars and galaxies (see pages 118 and 124).

Mind the gap...

There's some tricky stuff on this page, but it all boils down to "the bigger the gap, the less diffraction, the clearer the image". Sometimes diffraction is a good thing though — like when we want to split up light from stars to look at spectra. That proves to be really useful for finding out what stars are made of (see p.118).

Revision Summary for Section Nine

Well that's Section Nine done and dusted. What's the best way to celebrate? With some questions I reckon.

1)* What is the refractive index of a material if the speed of light through it is 1.25×10^8 m/s?

2)* Calculate the refractive index of a block of clear plastic if a beam of light enters it with an angle of incidence of 27° and is refracted at an angle of 18°.

3) What is dispersion?

4) a) What is total internal reflection?
b) What happens if the angle of incidence is less than the critical angle?
c) What happens if it is more than the critical angle?

5) Describe how total internal reflection is used in optical fibres.

6)* What is the critical angle of a beam of light hitting the boundary going from glass to air? The refractive index of air is 1 and the refractive index of glass is 1.52.

7) Describe the refraction of light by: a) a converging lens, b) a diverging lens.

8) What is meant by the principal focus of a lens?

9) Describe the differences between a real and a virtual image.

10) Describe the characteristics of an image formed from light from an object nearer to a converging lens than its principal focus, F.

11) What type of lenses are used to make magnifying glasses?

12)* Peter measures the length of a seed to be 1.5 cm. When he looks at the seed through a converging lens at a certain distance, the seed appears to have a length of 4.5 cm. What is the magnification of this lens at this distance?

13)* Find the power of a lens with a focal length of 0.25 m.

14) What two things affect the focal length of a lens?

15) Give the lens equation.

16)* An object was placed 0.2 m in front of a converging lens with a focal length of 0.15 m. How far behind the lens was the image formed?

17) Draw a simple sketch of the eye and label the following:
a) cornea, b) iris, c) pupil, d) lens, e) retina, f) ciliary muscles.

18) What is the far point of vision?

19) What is the near point of vision? Give an approximate value of the near point for adults.

20) Explain the symptoms and causes of: a) short sight, b) long sight.

21) What type of lens could be used to correct: a) short sight, b) long sight?

22) Describe how lasers can be used to correct vision problems.

23)* Ed Halley has a telescope with an $f_o = 10.0$ m and $f_e = 0.25$ m. What is the magnification of Ed's telescope?

24) Draw a labelled ray diagram for light from a distant object being focused in a concave mirror.

25) Why are concave mirrors used for astronomical telescopes?

26) Explain why a diffraction spectrum is created when white light passes through a diffraction grating.

*Answers on page 140.

The Earth's Structure

No one accepted the theory of plate tectonics for ages. Almost everyone does now. How times change.

The Earth Has a Crust, Mantle, Outer and Inner Core

The Earth is almost spherical and it has a layered structure, a bit like a scotch egg. Or a peach.

1) The bit we live on, the crust, is very thin (about 20 km).
2) Below that is the mantle. The mantle has all the properties of a solid, except that it can flow very slowly. Within the mantle, radioactive decay takes place (see p.72). This produces a lot of heat, which causes the mantle to flow in convection currents.
3) At the centre of the Earth is the core, which we think is made of iron and nickel. The inner core is solid but the outer core surrounding it is a liquid.

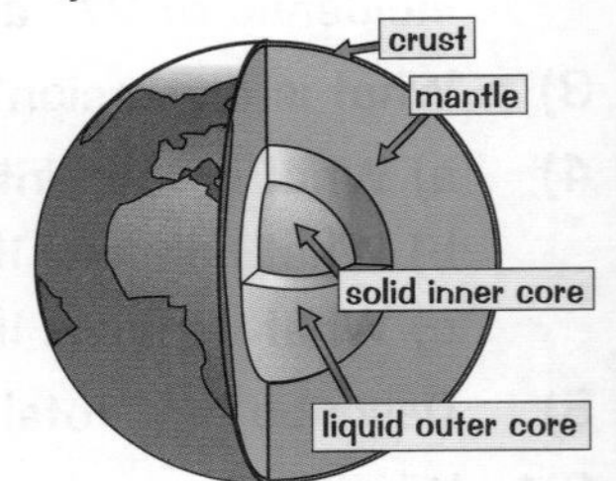

The Earth's Surface is Made Up of Tectonic Plates

1) The crust and the upper part of the mantle are cracked into a number of large pieces called tectonic plates. These plates are a bit like big rafts that 'float' on the mantle.
2) The plates don't stay in one place though. That's because the convection currents in the mantle cause the plates to drift.
3) Most of the plates are moving at speeds of a few cm per year relative to each other.
4) At plate boundaries, the plates may slide past each other — which sometimes causes earthquakes.
5) Volcanoes often form at the boundaries between two tectonic plates — magma (molten rock) is produced where plates meet, which can rise up and form volcanoes.
6) As plates crash into each other, mountains are formed. For example, the Himalayas are where India is crashing into the Eurasian plate.

Wegener Came Up with the Theory of Continental Drift

1) For years, very similar fossils had been found on opposite sides of the Atlantic Ocean.
2) Other things about the Earth also puzzled people — like why the coastlines of Africa and South America matched so well. And why fossils from sea creatures had been found high in the Alps.
3) Alfred Wegener hypothesised that Africa and South America had previously been one continent which had then split. Wegener's theory of 'continental drift' supposed that about 300 million years ago there had been just one 'supercontinent' — which he called Pangaea. According to Wegener, Pangaea broke into smaller chunks... and these chunks (our modern-day continents) are still slowly 'drifting' apart.
4) In the 1950s, scientists investigated the Mid-Atlantic ridge, which runs the whole length of the Atlantic.
5) They found evidence that magma rises up through the sea floor, solidifies and forms underwater mountains that are roughly symmetrical either side of the ridge. The evidence suggested that the sea floor was spreading — by a few cm per year.
6) Even better evidence that the continents are moving apart came from the magnetic orientation of the rocks. As the liquid magma erupts out of the gap, iron particles in the rocks tend to align themselves with the Earth's magnetic field — and as it cools they set in position. Now then... every half million years or so the Earth's magnetic field swaps direction — and the rock on either side of the ridge has bands of alternate magnetic polarity, symmetrical about the ridge.
7) This was convincing evidence that new sea floor was being created... and continents were moving apart.

Plate Tectonics — it's a smashing theory...

Some countries are particularly susceptible to earthquakes, which can also cause giant waves called tsunamis. Unfortunately, predicting them's very hard, and scientists aren't agreed on which method works best.

Seismic Waves

You can't drill very far into the crust of the Earth (only about 12 km), so scientists use seismic waves produced by earthquakes to investigate the Earth's inner structure.

Earthquakes Cause Different Types of Seismic Waves

1) When there's an earthquake, it produces wave motions (shock waves) which travel on the surface and inside the Earth. We record these seismic waves all over the surface of the planet using seismographs.
2) Seismologists measure the time it takes for the shock waves to reach each seismograph.
3) They also note which parts of the Earth don't receive the shock waves at all.
4) There are two different types of seismic waves that travel through the Earth — P-waves and S-waves.

P-Waves Travel Through Solids and Liquids

1) P-waves travel through solids and liquids.
2) They travel faster than S-waves.
3) P-waves are longitudinal (see p.27).

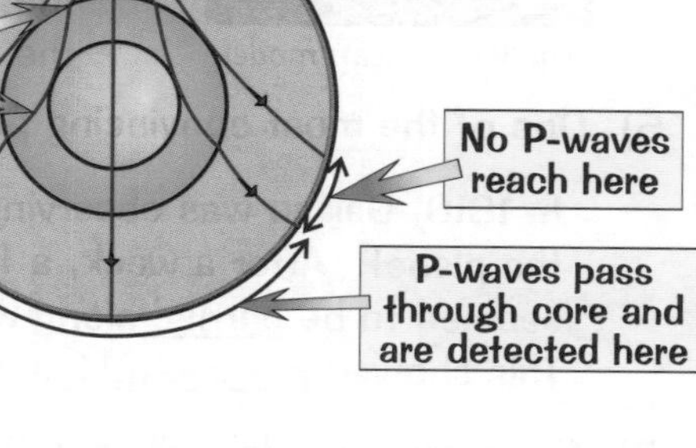

S-Waves Only Travel Through Solids

1) S-waves only travel through Solids.
2) They are Slower than P-waves.
3) S-waves are transverse (see p.27).

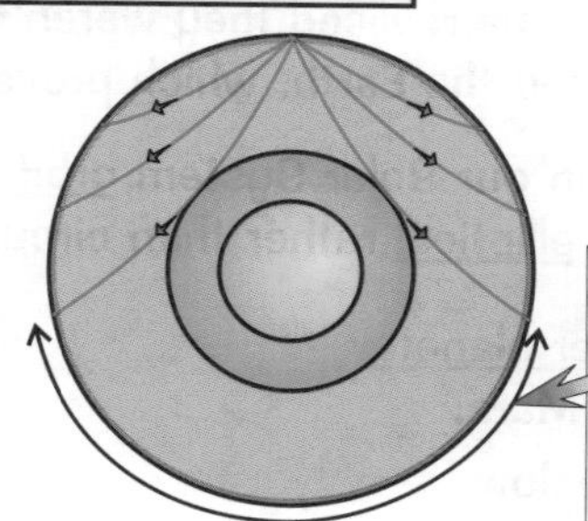

There's also a third type of wave — surface waves. But you don't need to know about those.

The Waves Curve with Increasing Depth

1) When seismic waves reach a boundary between different layers of the Earth, some waves will be reflected.
2) The waves also change speed as the properties (e.g. density) of the mantle and core change. This change in speed causes the waves to change direction — which is refraction.
3) Most of the time the waves change speed gradually, resulting in a curved path. But when the properties change suddenly, the wave speed changes abruptly, and the path has a kink.

The Seismograph Results Tell Us What's Down There

1) About halfway through the Earth, P-waves change direction abruptly. This indicates that there's a sudden change in properties — as you go from the mantle to the core.
2) The fact that S-waves are not detected in the core's shadow tells us that the outer core is liquid — S waves only pass through Solids.
3) P-waves seem to travel slightly faster through the middle of the core, which strongly suggests that there's a solid inner core.
4) Note that S-waves do travel through the mantle, which shows that it's solid. It only melts to form magma in small 'hot spots'.

What's that coming straight through the core? Is it a P-wave, is it a P-wave?

You need to remember that P-waves are longitudinal and S-waves are transverse. You might find it helpful to think of them as Push-waves and Shake-waves. Gosh — what a useful little trick. You can thank me later...

The Solar System

Our Solar System is made up of a star (the Sun) and lots of stuff orbiting it in slightly elongated circles (called ellipses). But scientists and astronomers didn't always think the Universe and our Solar System was like that...

Ancient Greeks Thought the Earth was the Centre of the Universe

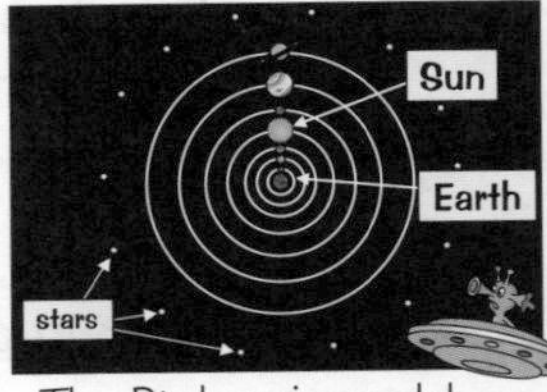

The Ptolemaic model.

1) Most ancient Greek astronomers believed that the Sun, Moon, planets and stars all orbited the Earth in perfect circles — this is the geocentric or Ptolemaic model.
2) The Ptolemaic model was the accepted model of the Universe until the 1600s, when it began to be replaced by the heliocentric or Copernican model.

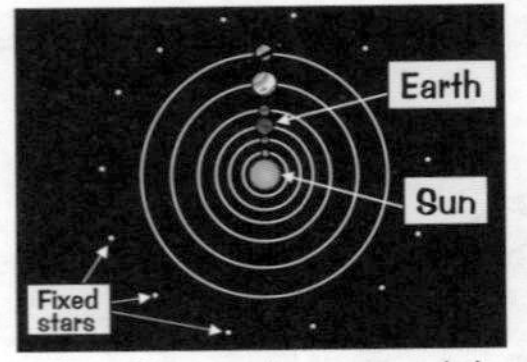

The Copernican model.

3) The Copernican model states that the Earth and planets all orbit the Sun, which is at the centre of the Universe.
4) The heliocentric model was first introduced in a book by Copernicus in 1543. This book showed astronomical observations could be explained without having the Earth at the centre of the Universe. Copernicus' ideas weren't popular at the time, and the model itself was condemned by the Church.
5) One of the most convincing pieces of evidence for the theory was Galileo's observations of Jupiter's moons.

In 1610, Galileo was observing Jupiter using a telescope (a new invention) when he saw three stars near the planet. After a week, a fourth star appeared. These stars never moved away from Jupiter and seemed to be carried along with the planet — he realised they weren't stars, but moons orbiting Jupiter. This showed not everything was in orbit around the Earth, which proved the geocentric model was wrong.

6) The current model still says that the planets in our Solar System orbit the Sun — but that these orbits are actually elliptical rather than circular.

- Closest to the Sun are the inner planets — Mercury, Venus, Earth and Mars.
- Then the asteroid belt — see below.
- Then the outer planets, much further away — Jupiter, Saturn, Uranus, Neptune.

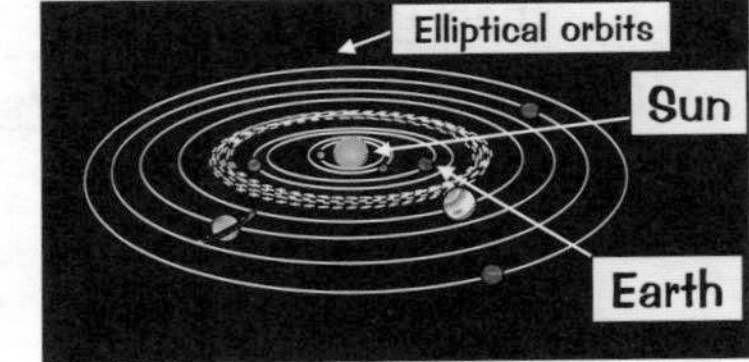

Our current view of the Solar System.

7) As technology has improved, our idea of the Solar System and the Universe has changed. E.g. the invention of the telescope led to the discovery of Uranus.

Planets Reflect Sunlight

1) You can see some planets with the naked eye. They look like stars, but they're totally different.
2) Stars are huge, very hot and very far away. They give out lots of light — which is why you can see them, even though they're very far away.
3) The planets are smaller and nearer and they just reflect sunlight falling on them.
4) Planets often have moons orbiting around them. Jupiter has at least 63 of 'em. We've just got one.

Asteroids and Comets are Smaller Than Most Planets

1) Asteroids and comets are made of stuff left over from the formation of the Solar System.
2) The rocks between Mars and Jupiter didn't form a planet, but stayed as smallish lumps of rubble and rock — these are asteroids.
3) Comets are balls of rock, dust and ice which orbit the Sun in very elongated ellipses, often in different planes from the planets. The Sun is near one end of the orbit.
4) As a comet approaches the Sun, its ice melts, leaving a bright tail of gas and debris which can be millions of kilometres long. This is what we see from Earth.

Asteroids... my dad had those — very nasty...

It's taken thousands of years to get our current model of the Solar System, and there's loads we don't know...

Beyond the Solar System

There's all sorts of exciting stuff out there. The whole Solar System is just part of one galaxy. And there are billions upon billions of galaxies. Yup, the Universe is big — huge in fact...

We're in the Milky Way Galaxy

1) Our Sun is one of thousands of millions of stars which form the Milky Way galaxy — about 1 in 100 000 000 000 (or 10^{11}) if you had to write it out.
2) The Sun is about halfway along one of the spiral arms of the Milky Way.
3) The distance between neighbouring stars in a galaxy is usually millions of times greater than the distance between planets in the Solar System.

The Whole Universe Has More Than a Thousand Million Galaxies

1) Every galaxy is made up of thousands of millions of stars, and the Universe is made up of thousands of millions of galaxies — that's a lot of stars.
2) Galaxies themselves are often millions of times further apart than the stars are within a galaxy.
3) So even the slowest among you will have worked out that the Universe is mostly empty space and is really really BIG.

Distances in Space Can Be Measured Using Light Years

1) Once you get outside our Solar System, the distances between stars and between galaxies are so enormous that kilometres seem too pathetically small for measuring them.
2) For example, the closest star to us (after the Sun) is about 40 000 000 000 000 kilometres away (give or take a few hundred billion kilometres). Numbers like that soon get out of hand.
3) So we use light years instead. A light year is the distance that light travels through a vacuum (like space) in one year. Simple as that.
4) Light travels really fast — 300 000 km/s. So 1 light year is equal to about 9 460 000 000 000 km.
5) Just remember — a light year is a measure of DISTANCE (not time).

You Need to Know Some Relative Sizes, Distances and Ages

You need to know the relative sizes and distances of some different stuff in space — this just means the size in relation to the size of something else, e.g. the Earth's diameter is smaller than the Sun's:

SMALLEST

- Diameter of the Earth
- Diameter of the Sun
- Diameter of the Earth's orbit
- Diameter of the Solar System
- Distance from the Sun to the nearest star
- Diameter of the Milky Way
- Distance from the Milky Way to the nearest galaxy

LARGEST

1) The Sun's diameter is about 100 times bigger than the diameter of the Earth.
2) The diameter of the Milky Way is about 600 billion times the diameter of the Sun. Yup... it's pretty big.
3) The Milky Way and its nearest galaxy are about 600 000 times further apart than the Sun and its nearest star.

You need to know some ages too:

Different stuff in space	Age (million years)
Earth	5000
Sun	5000
Universe	14 000

The Earth and the Sun are a similar age. But the Universe is about 3 times older.

You may think it's a long way down the street to the chip shop...

...but that's nothing compared to distances in space. Space is also less tasty (and even worse for your health).

The Life Cycle of Stars

Stars go through many traumatic stages in their lives — just like teenagers.

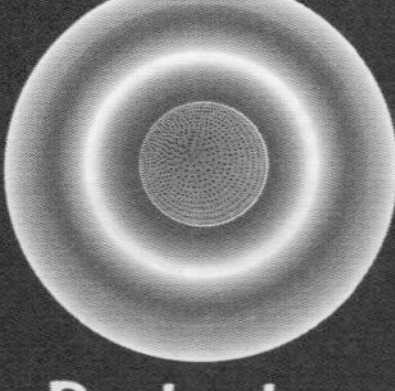

1) Stars initially form from clouds of DUST AND GAS. The force of gravity makes the gas and dust spiral in together to form a protostar.

Protostar

2) Gravitational energy is converted into heat energy, so the temperature rises. When the temperature gets high enough, hydrogen nuclei undergo nuclear fusion to form helium nuclei and give out massive amounts of heat and light. A star is born. Smaller masses of gas and dust may also pull together to make planets that orbit the star.

Main Sequence Star

3) The star immediately enters a long stable period, where the heat created by the nuclear fusion provides an outward pressure to balance the force of gravity pulling everything inwards. The star maintains its energy output for millions of years due to the massive amounts of hydrogen it consumes. In this stable period it's called a MAIN SEQUENCE STAR and it lasts several billion years. (The Sun is in the middle of this stable period — or to put it another way, the Earth has already had half its innings before the Sun engulfs it!)

Stars much bigger than the Sun

Stars about the same size as the Sun

4) Eventually the hydrogen begins to run out. Heavier elements are made by nuclear fusion of helium. The star then swells into a RED GIANT, if it's a small star, or a RED SUPER GIANT if it's a big star. It becomes red because the surface cools.

Red Giant

White Dwarf

Red Super Giant

5) A small-to-medium-sized star like the Sun then becomes unstable and ejects its outer layer of dust and gas as a PLANETARY NEBULA.

6) This leaves behind a hot, dense solid core — a WHITE DWARF, which just cools down to a BLACK DWARF and eventually disappears.

Supernova

Neutron Star...

...or Black Hole

7) Big stars, however, start to glow brightly again as they undergo more fusion and expand and contract several times, forming elements as heavy as iron in various nuclear reactions. Eventually they explode in a SUPERNOVA, forming elements heavier than iron and ejecting them into the Universe to form new planets and stars.

8) The exploding supernova throws the outer layers of dust and gas into space, leaving a very dense core called a NEUTRON STAR. If the star is big enough this will become a BLACK HOLE.

Red Giants, White Dwarfs, Black Holes, Green Ghosts...

The early Universe contained only hydrogen, the simplest and lightest element. It's only thanks to nuclear fusion inside stars that we have any of the other naturally occurring elements. Remember — the heaviest element produced in stable stars is iron, but it takes a supernova (or a lab) to create the rest.

The Life Cycle of Stars

Stars — without them it would be really, really cold (and dark, and boring) in the Universe. Hooray for stars.

Fusion Happens in the Core of a Star

1) A star is made up of a core (centre) surrounded by different layers.
2) The closer to the centre of the star, the hotter that bit will be, e.g. the core is hotter than the surface.

THE CORE — Most of the fusion in a star takes place in the centre. The pressure from the weight of the rest of the star makes the core hotter and denser than the rest of the star. So, the nuclei in the core are close enough (and have enough energy) to fuse together.

SURFACE (photosphere) — the outer region of the star, from where energy is radiated into space. Energy released from fusion in the core is transported by radiation and convection currents to the surface of the star. This is the part of the Sun that we see from Earth.

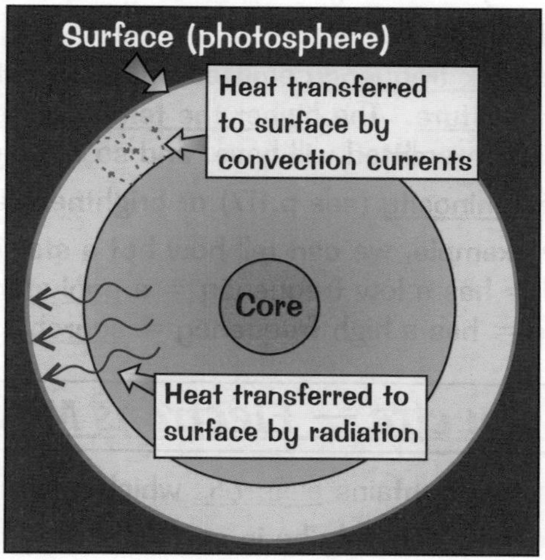

Luminosity vs Temperature — the Hertzsprung-Russell Diagram

1) If you plot luminosity (brightness) against temperature, you don't just get a random collection of stars.
2) Different types of stars group together in distinct areas. This is called the Hertzsprung-Russell diagram.
3) The different areas show the main stages of a star's life cycle: the main sequence, red giants and super giants and white dwarfs.

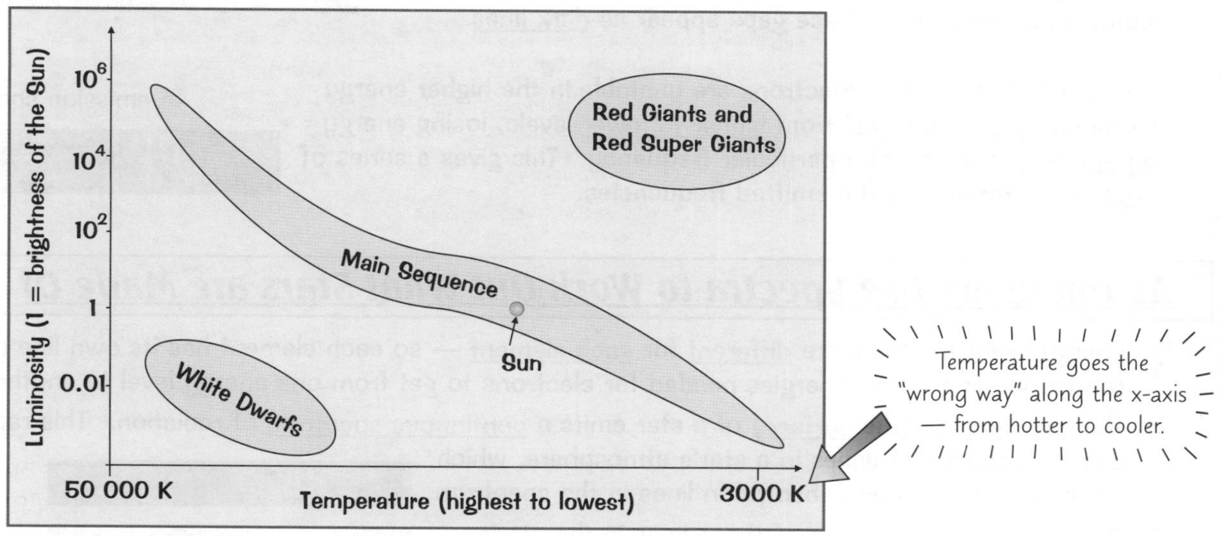

4) The reason you can see these areas is because stars exist in these stable stages of their life cycle for long periods of time.
5) You don't see unstable phases, like supernovas, on the diagram because they happen too quickly.

I bet all stars wish they could keep their unstable phases hidden...

I'm going to name my dog after Ejnar Hertzsprung, because I like his diagram — and Ejnar is just a great name.

Star Spectra

The light from stars doesn't just make the sky look pretty, it can actually tell you what stars are made of...

Continuous Spectra Contain All Possible Frequencies

1) All hot objects like stars emit radiation. Hot objects emit a continuous range of frequencies — a continuous spectrum (it doesn't have any gaps).
2) Hot objects always emit more of one frequency than any other. This wavelength is called the peak frequency.

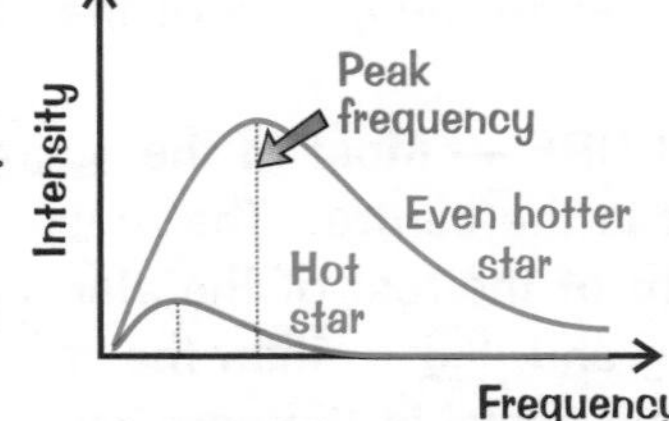

3) The peak frequency emitted by an object depends on its temperature. The higher the temperature, the more energy the radiation emitted will have, and so the higher the peak frequency.
4) The luminosity (see p.117) or brightness also depends on temperature — hotter things glow more.
5) For example, we can tell how hot a star is by looking at its colour:
   red = has a low frequency = a cool star (well... still hot enough to cook you and your toast)
   blue = has a high frequency = scorchio (that's hot). The Sun's a fairly cool star — it looks yellowy.

Line Spectra — Electrons Moving Between Energy Levels

1) An atom contains electrons which move around a tiny positive nucleus.
2) Electrons can only be in certain energy levels (or shells) around the nucleus — with the lowest energy levels nearest the nucleus.
3) Electrons move between energy levels if they gain or lose energy.
4) Electrons can gain enough energy to be removed from the atom — this is called ionisation (p.35).

energy levels

ABSORPTION SPECTRA — At high temperatures, electrons become excited and jump into higher energy levels by absorbing radiation. Because there are only certain energy levels an electron can occupy, electrons absorb a particular frequency of radiation to get to a higher energy level.
You can 'see' this if a continuous spectrum of visible light shines through a gas — the electrons in the gas atoms absorb certain frequencies of the light, making gaps in the otherwise continuous spectrum. These gaps appear as dark lines.

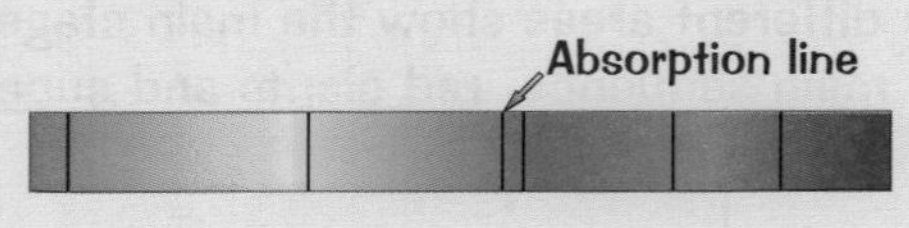

EMISSION SPECTRA — Electrons are unstable in the higher energy levels so they tend to fall from higher to lower levels, losing energy by emitting radiation of a particular frequency. This gives a series of bright lines formed by the emitted frequencies.

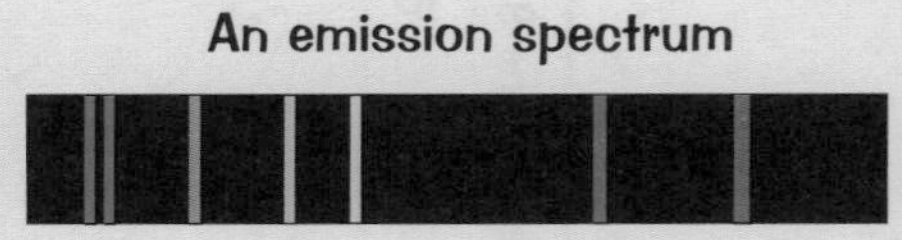

Astronomers Use Spectra to Work Out What Stars are Made Of

1) Energy levels in atoms are different for each element — so each element has its own line spectrum (corresponding to the energies needed for electrons to get from one energy level to another).
2) The photosphere (the surface) of a star emits a continuous spectrum of radiation. This radiation passes through the gases in a star's atmosphere, which produces emission and absorption lines in the spectrum.
3) By looking at the position of these lines in the star's spectrum, you can work out what chemical elements are present in the star's atmosphere — by comparing it with known spectra in the lab.

Stellar spectrum (containing H, He and Na)
Hydrogen
Helium
Sodium

What do you call a star detective? In-Spectra...

It's really useful stuff this line spectra business — it's as near as you can get to getting a spaceship, flying to a star, getting out your bucket and spade and having a good dig around to see what it's made of. Pretty neat, huh.

Observing the Sky

It's easy to see why people thought we were at the centre of the Universe for ages.
The Sun, Moon and stars all seem to orbit around us — but really it's all down to the Earth's spin.

A Sidereal Day is the Time Taken for the Earth to Spin Once

1) If you looked at the night sky for long enough, you'd see distant stars appear to cross the sky from east to west.
2) Astronomers have known for years that it's not the stars that move, but the Earth that spins on its axis.
3) For a star to get to the same position in the sky, the Earth needs to spin 360°. The time taken for this to happen is called a sidereal day.

A sidereal day is the time taken for a star to return to the same position in the sky. It's about 23 hours and 56 minutes.

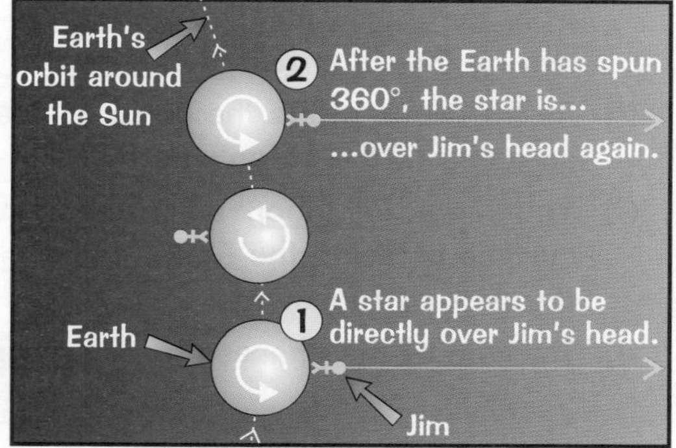

The Sun and Moon Appear to Cross the Sky at Different Speeds

1) It's not just the stars — the Sun and Moon also appear to cross the sky from east to west.
2) The Sun seems to move more slowly across the sky than distant stars — it takes 24 hours to get to the same position in the sky, a whole 4 minutes longer. This is called a solar day.

A solar day is the time taken for the Sun to appear at the same position in the sky. It's 24 hours.

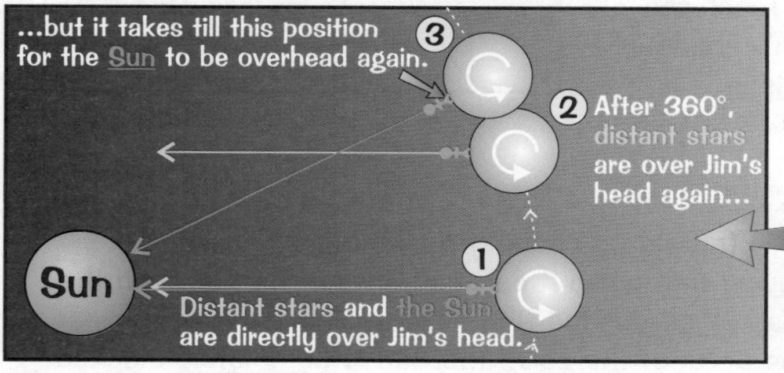

3) Solar and sidereal days are different because the Earth orbits the Sun as well as spinning on its axis.
4) The Earth orbits the Sun in the same direction as it spins — so the Earth needs to spin slightly more than 360° before the Sun appears at the same position in the sky.

5) The Moon seems to go more slowly than the Sun, taking about 25 hours to appear at the same position in the sky.
6) This is because the Moon orbits the Earth in the same direction as the Earth is rotating.

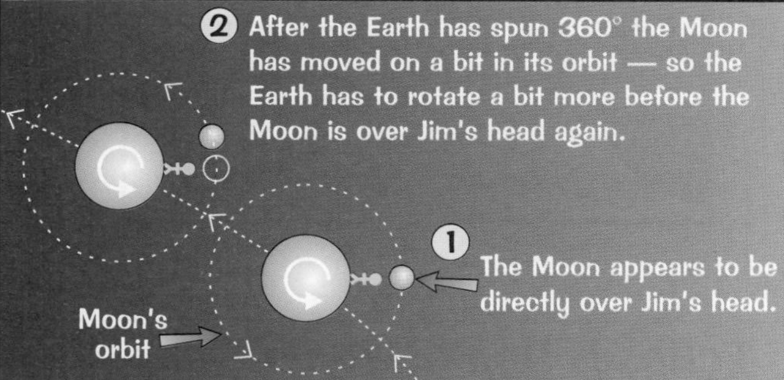

The Stars You Can See in the Sky Change During the Year

1) As the Earth moves around the Sun, the direction we face changes slightly each day.
2) This means we can see a slightly different patch of sky each night — we see different stars.
3) An Earth year is the time it takes the Earth to orbit the Sun once, so on the same day each year you should be able to see the same stars in the night sky.

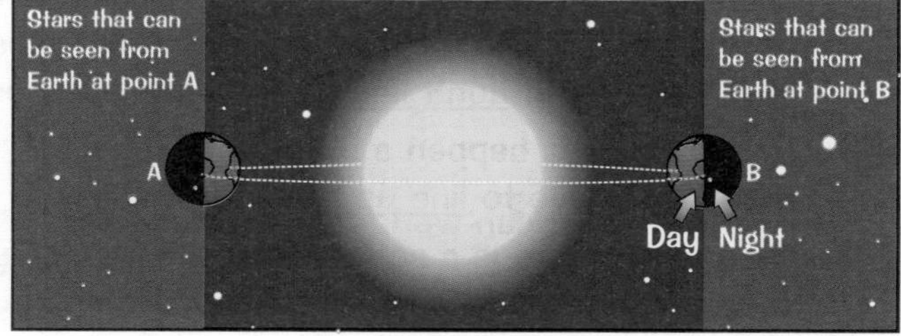

All this spinning is making me dizzy...

The ancient Greeks thought the Sun was a god called Helios riding a golden chariot across the sky. Bit fancier than your average car I suppose. Don't try putting that in your exam though... you won't get any marks.

Eclipses and the Moon

Some more spinning and orbiting, but this time it's the Moon's turn...

The Phases of the Moon

1) The Moon doesn't glow itself — it only reflects light from the Sun. Only the half facing the Sun is lit up, leaving the other half in shadow.
2) As the Moon orbits the Earth, we see different amounts of the Moon's dark and lit-up surfaces.
3) You see a 'full moon' when the whole of the lit-up surface is facing the Earth, and a 'new moon' when the dark half faces us.
4) The rest of the phases are in between these two extremes.

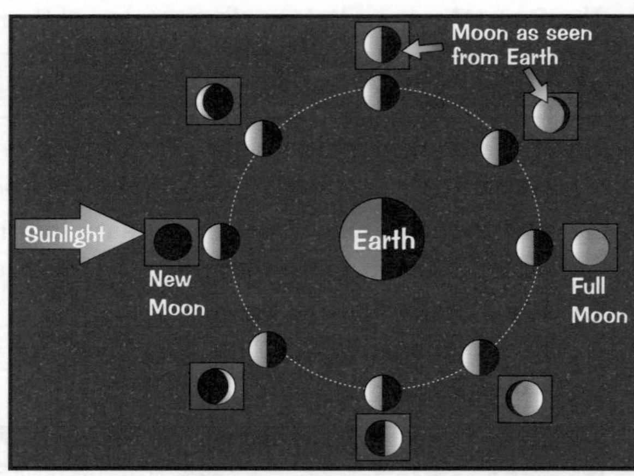

Eclipses Happen When Light from the Sun is Blocked

There are two main types of eclipse: lunar and solar.

LUNAR ECLIPSE

As it orbits, the Moon sometimes passes into the Earth's shadow. The Earth blocks sunlight from the Moon, so almost no light is reflected from the Moon and it just seems to disappear. A total lunar eclipse is where no direct sunlight can reach the Moon. More often, the Moon isn't fully in the Earth's shadow so only part of it appears dark — a partial lunar eclipse.

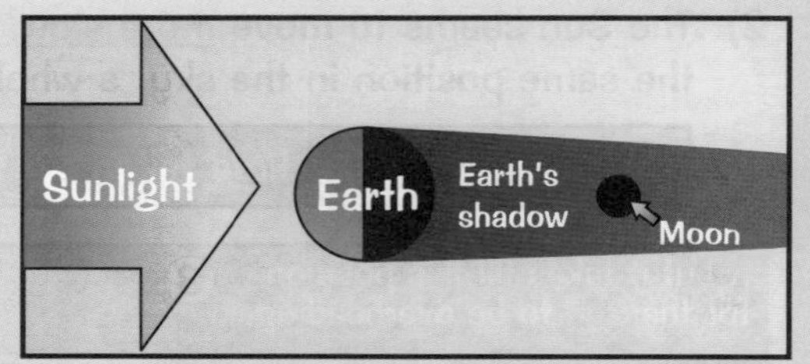

SOLAR ECLIPSE

Sun
Total eclipse
Earth
Moon
Partial eclipse

The Moon is (purely by chance) just the right size and distance away that when it passes between the Sun and the Earth, it can block out the Sun. This is called a solar eclipse. From some parts of the Earth the Sun is completely blocked — a total solar eclipse. From many places on Earth only part of the Sun will be blocked — a partial solar eclipse. From most places on Earth the Sun won't be blocked at all.

Eclipses Don't Happen Very Often

1) It's not every day you see the Sun being blotted out of the sky by the Moon.
2) The Moon orbits the Earth at an angle to Earth's orbit around the Sun. So most of the time the Sun, Moon and Earth don't line up to cause eclipses.
3) Partial eclipses happen a bit more often as they don't have to line up perfectly for this.

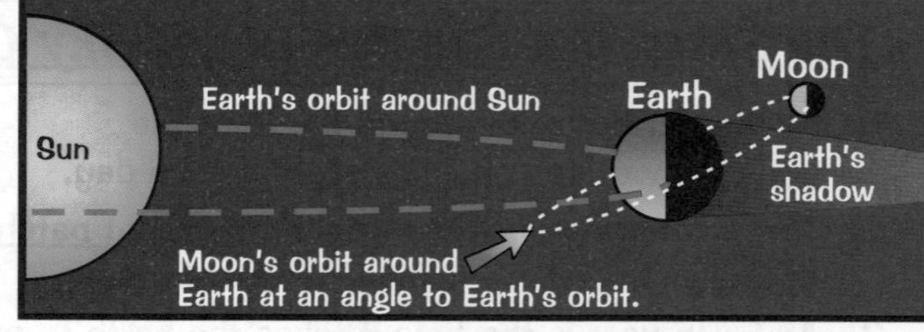

4) Even when there is a solar eclipse, there's only a very small region on Earth from which it can be seen. There might be a total solar eclipse in China but we'd hardly notice anything in the UK.

Aaaargh — the Sun's been eaten by a giant sky monster...

If you ever get the chance to see a total solar eclipse, it's amazing (but don't look directly at the Sun). If you want to see it in the UK though, don't buy your popcorn just yet — you'll have to wait 'til 2090...

Coordinates in Astronomy

This page is a bit tricky, but stick with it...

The Positions of Stars are Measured by Angles Seen from Earth

1) The positions of stars are measured by angles seen from Earth. It's just like latitude and longitude on Earth — only for the sky.
2) The sky appears to turn as the Earth spins — so astronomers picked two fixed positions to measure from:

   THE POLE STAR is a star that doesn't seem to move because it's almost directly above the North Pole (and the spin axis) of Earth.

   THE CELESTIAL EQUATOR is an imaginary plane running across the sky, extending out from the Earth's equator.
3) The two angles used to measure positions in the sky are:

   Declination — Celestial latitude, measured in degrees.

   Right Ascension — Celestial longitude (the 'how much across' angle), measured in degrees or time.

   It might sound a bit weird having an angle as a time, but it's possible because the Earth turns through 360° every 24 hours. Right ascension increases the further east on the sky you go.

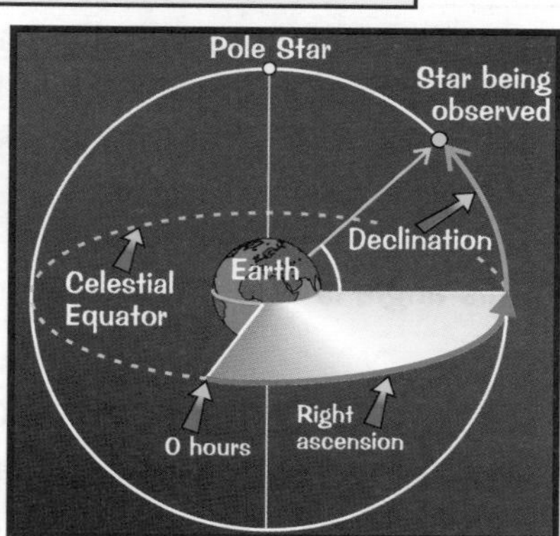

Planets Seem to Move in Complicated Patterns

1) All the planets in the Solar System orbit the Sun in the same direction but at different speeds. The closer to the Sun, the quicker the planet.
2) Even without a telescope, you can often see the 'naked eye' planets — Mercury, Venus, Mars, Jupiter and Saturn. To track a planet as it goes across the sky, set your alarm clock and note down its position at the same (sidereal) time each night. (That way you rule out the Earth's spin making it cross the sky.) You'll find that the planets seem to gradually travel from west to east.
3) Every so often, though, a planet seems to change direction and go the other way for a bit (in relation to the fixed stars), making a loop or squiggle in its track before carrying on as normal. This is called retrograde motion. It only happens with the outer planets — Mars to Neptune.
4) It happens because both the planet and Earth are moving around the Sun — so we're seeing the motion of the planet relative to Earth.
5) Mars appears to change direction once every two or so years. Slower-moving planets further out 'change direction' less frequently.

Fixed stars are distant background stars that stay in the same position year after year.

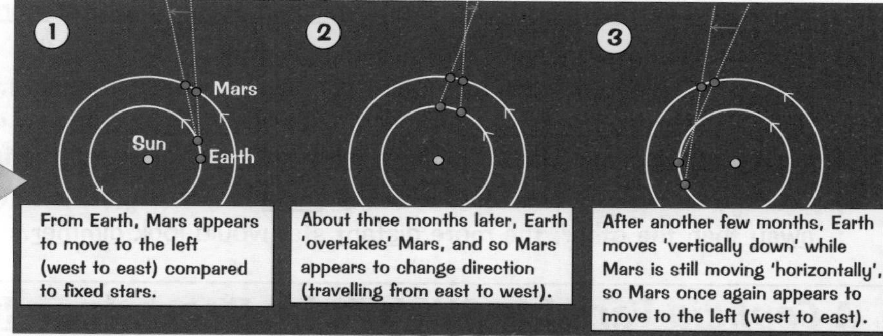

1: From Earth, Mars appears to move to the left (west to east) compared to fixed stars.

2: About three months later, Earth 'overtakes' Mars, and so Mars appears to change direction (travelling from east to west).

3: After another few months, Earth moves 'vertically down' while Mars is still moving 'horizontally', so Mars once again appears to move to the left (west to east).

Date	Right Ascension		Declination (degrees)
	hours	minutes	
June 5th 2003	21	51	-16.3
July 31st 2003	22	55	-14.6
Sept 10th 2003	22	45	-16.5
Oct 20th 2003	22	25	-13.2
Dec 1st 2003	23	29	-4.3

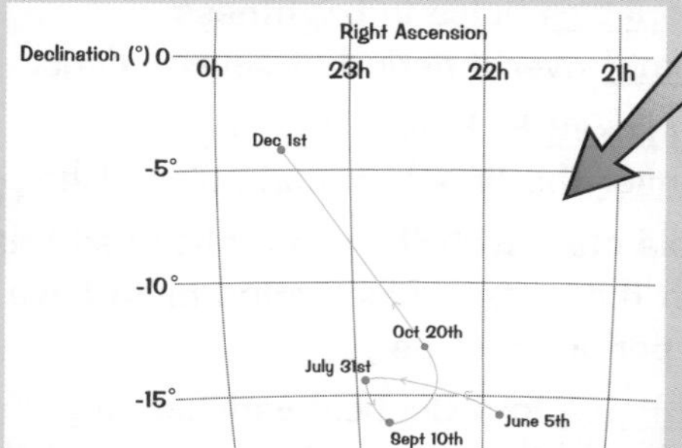

The diagram shows the loop Mars made in 2003. You can tell just from the table of data that Mars has changed direction. Mars travels east (right ascension increases) — then changes direction and moves west (right ascension decreases) before moving east again (right ascension increases).

It's all just loops and squiggles...

Time is an angle? Hmm. And the planets are a bit like athletes running at different speeds on a race track and lapping each other — make sure you can explain how this causes those loops and squiggles.

Astronomical Distances and Brightness

Stars — they're bright and really far away. But are the brighter ones really brighter or just closer...

The Distance to Nearby Stars can be Measured by Parallax

1) Parallax is an apparent change in position of an object against a distant background. It makes closer stars appear to move relative to distant ones over the course of a year. In astronomy:

> The parallax angle is half the angle moved against distant background stars over 6 months (at the opposite ends of the Earth's orbit). The nearer an object is to you, the greater the angle.

2) This angle is often measured in arcseconds (or 'seconds of arc') rather than degrees:

$$1 \text{ arcsecond} = 1'' = \left(\frac{1}{3600}\right)^{\circ}$$

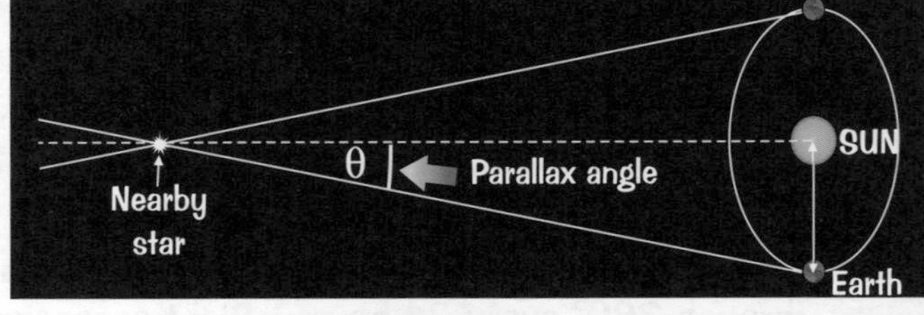

3) Parallax is useful for calculating the distance to nearby stars. The smaller the parallax angle, the more distant the star is.
4) Astronomers usually use a unit of distance called a parsec. It's a similar magnitude to a light year (1 parsec = about 3 light years) — a really long way. Distances between stars (interstellar distances) are normally a few parsecs.

> A parsec (pc) is the distance to a star with a parallax angle of 1 arcsecond.

5) You can calculate the distance to a star (in parsecs) using this equation:

$$\text{distance (pc)} = \left(\frac{1}{\text{angle (arcsec)}}\right)$$

EXAMPLE: Mr Moore measures the parallax of a star to be 0.4''. How far away is the star in parsecs?

ANSWER: $\text{distance (pc)} = \left(\frac{1}{0.4''}\right) = \underline{2.5 \text{ pc}}$

Observed Intensity Depends on the Distance to the Star

1) The luminosity, or intrinsic brightness, of a star (how bright it'd seem if you were right next to it) depends on its size and temperature. The bigger and hotter it is, the more energy it gives out, so the brighter it is.
2) As you move away from a star, it looks dimmer — because the energy reaching you gets less (as it spreads out through space). So the observed intensity of the light (the observed brightness) of a star as seen on Earth depends on its luminosity and how far away it is from Earth.
3) So if you looked at two stars with the same luminosity but one was further away than the other, the more distant star would look dimmer.

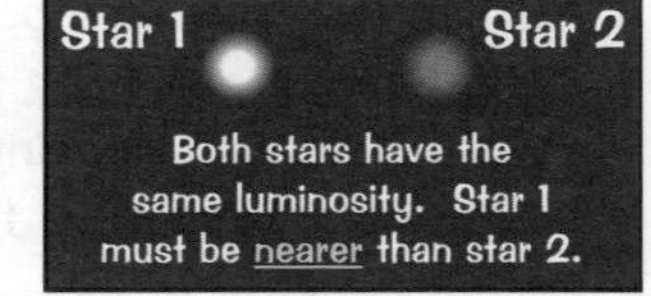

Both stars have the same luminosity. Star 1 must be nearer than star 2.

A Cepheid Variable Star's Pulse Depends on Its Luminosity

1) A group of stars called Cepheid variables pulse in brightness — they get brighter and then dimmer over a period of several days.
2) How quickly they pulse is directly linked to their luminosity. The greater the luminosity, the longer the time between pulses (the pulse period).
3) So, if you see two Cepheid variable stars with the same observed brightness that pulse at different rates, you know that the star with the longer pulse period must have the higher luminosity. So the star with the longer pulse period must be further away.
4) Astronomers can work out the distance to a Cepheid variable by comparing the luminosity (worked out from the pulse period) and the observed brightness of the star.

I'm not dim — I'm just really far away...

Cepheid variables are really useful — you can use them to work out the distance to whatever big ball of gas they might be in. These cheeky chaps helped solve one of the most famous debates in astronomy... Next page...

The Scale of the Universe

We mightn't know exactly how big, but we know the Universe is biiiiiiiiiiiiiig.

The Curtis-Shapley Debate

In the 1920s there was a debate about the size and structure of the Universe, led by two famous American astronomers — Harlow Shapley and Heber Curtis. Through telescopes, people had seen some faint, fuzzy objects that they called nebulae. Some of these objects looked spiral-shaped but some were just blobs. Shapley and Curtis argued about what these nebulae were and where they were:

Shapley's Argument

1) Shapley believed the Universe was just one gigantic galaxy about 100 000 parsecs across.
2) He reckoned our Sun and Solar System were far from the centre of the galaxy.
3) He believed that nebulae were huge clouds of gas and dust. These clouds were relatively nearby and actually part of the Milky Way.

Curtis' Argument

1) Curtis thought the Universe was made up of many galaxies.
2) He thought our galaxy was smaller than Shapley suggested — about 10 000 pc across, with the Sun at or very near the centre.
3) The spiral nebulae were other very distant galaxies, completely separate from the Milky Way.

And the winner is... well both of them really:

1) Shapley was right that the Solar System is far from the centre of our galaxy, but Curtis was right that there are many galaxies in the Universe (at least 100 000 000 000 of them).
2) Curtis was also right about spiral nebulae — Hubble used Cepheid variable stars to show that they're really far away (see below). (The debate wasn't properly over until the 1930s, though, when better telescopes meant we could see the nebulae clearly, rather than just as blurry blobs.)

Hubble Showed There Were Objects Outside the Galaxy

1) Hubble helped solve the Curtis-Shapley debate with his observations of the Andromeda nebula.
2) Using images taken using the largest telescope at the time, he found that this spiral-shaped fuzzy blob contained many stars, some of which were Cepheid variables (see p.122).
3) Hubble calculated the distance to the Andromeda nebula by working out the distance to the Cepheid variables within it, using the relationship between their brightness and pulse frequency (see p.122).
4) He found it was about 2.5 million light years away — much further than any stars in our galaxy.
5) He studied other spiral nebulae and found a similar result — they were all too far away to be part of the Milky Way, and so must be separate spiral galaxies themselves.

The Milky Way — not just a tasty snack...

At least it was all sorted in the end, and now it's a joy for you to learn instead... well, maybe not.

The Origin of the Universe

How did the Universe begin? Why are we here? Do fish sleep? Yep, there are a lot of big brain curdling questions out there. Thankfully, physics has some clues about the answer to that first one... go physics!

The Universe Seems to be Expanding

As big as the Universe already is, it looks like it's getting even bigger.
All its galaxies seem to be moving away from each other. There's good evidence for this...

Light from Other Galaxies is Red-shifted

1) Different chemical elements absorb different frequencies (p.118) of light.
2) Each element produces a specific pattern of dark lines at the frequencies that it absorbs in the visible spectrum.
3) When we look at light from distant galaxies we can see the same patterns but at slightly lower frequencies than they should be — they're shifted towards the red end of the spectrum. This is called red-shift.

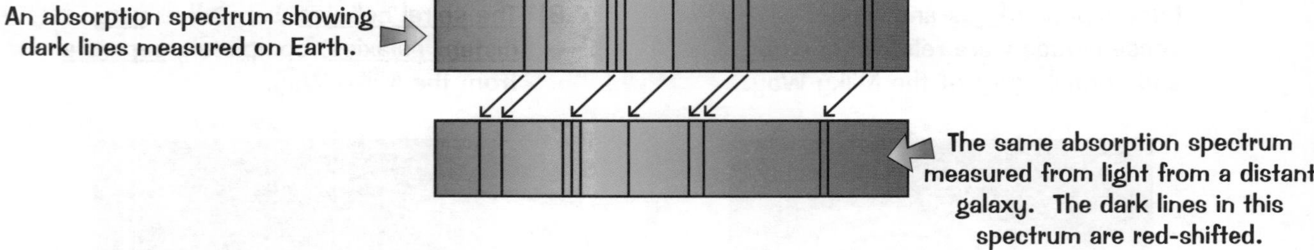

4) It's the same effect as the vrrroomm from a racing car — the engine sounds lower-pitched when the car's gone past you and is moving away from you. This is called the Doppler effect:

THE DOPPLER EFFECT

1) The frequency of a source moving towards you will seem higher and its wavelength will seem shorter.
2) The frequency of a source moving away from you will seem lower and its wavelength will seem longer.

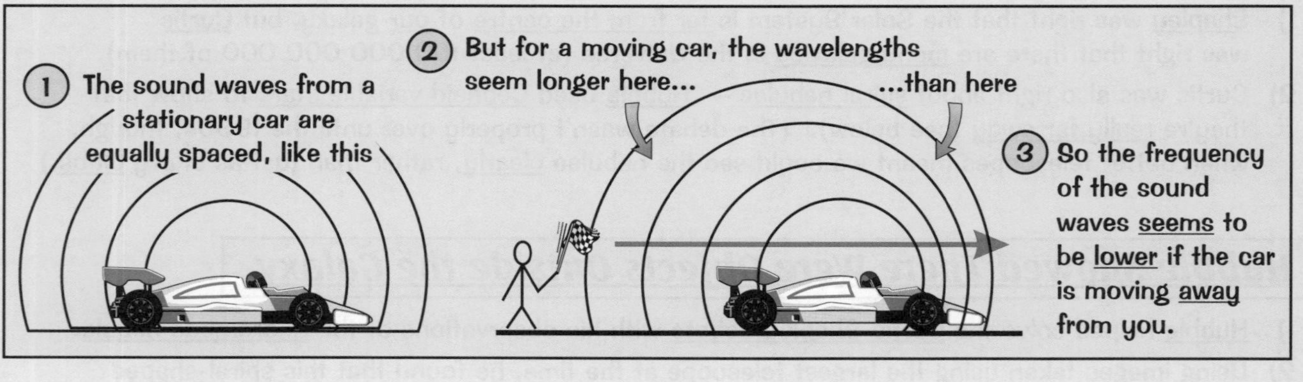

5) The faster a galaxy is moving away from us, the further the light from it will be red-shifted.
6) Measurements of the red-shift suggest that all the galaxies are moving away from us very quickly — and it's the same result whichever direction you look in.

If a tree falls down in the forest and you're driving away from it...

Listen out for the Doppler effect next time you hear a fast motorbike or a police siren — physics in action. Red-shift might not sound like the most thrilling thing in the world, but what the red-shifts of distant galaxies can tell you about the Universe is pretty awesome. More on that coming up...

The Origin of the Universe

Edwin Hubble was possibly the most famous astronomer ever (apart from our Patrick with his xylophone).

Distant Galaxies are Moving Away from Us

1) By seeing how much light from a galaxy has been red-shifted, you can work out the recession velocity of the galaxy (how quickly it's moving away). The greater the red-shift, the greater the speed of recession.
2) Using red-shift, Hubble compared the speed and the distances for many distant galaxies and found a pattern:

The more distant the galaxy, the faster it moves away from us.

3) Red shift is fairly easy to measure, so a galaxy's recession velocity can be calculated easily enough. The distance to a distant galaxy can be found from its recession velocity using Hubble's law.

Speed of recession	= Hubble constant	× distance
(km/s)	(s^{-1})	(km)
or (km/s)	(km/s per Mpc)	(Mpc)

A megaparsec (Mpc) is just a unit of distance. 1 Mpc is about 3×10^{19} km.

4) The Hubble constant can have the units s^{-1} or km/s per Mpc (depending on the unit of distance you use). The value of the Hubble constant is roughly 2×10^{-18} s^{-1} or 70 km/s per Mpc.
5) You could be asked to use the equation in the exam to find any of the variables...

EXAMPLE: Find the distance to a galaxy that has a recession velocity of 475 km/s.

ANSWER: Rearrange the equation: Distance = Speed of recession / Hubble constant
$= 475 \div (2 \times 10^{-18}) = 2.375 \times 10^{20}$ km.

6) The value of the Hubble constant is still being researched and there are many different ways to calculate it, unfortunately the different methods all come up with slightly different answers.
7) Using data on Cepheid variable stars from distant galaxies has given us better values of Hubble's constant.

It All Started Off with a Very Big Bang (Probably)

So, distant galaxies are moving away from us — the further away a galaxy is from the us, the faster they're moving away (see above). But something must have got them going. That 'something' was probably a big explosion — so they called it the Big Bang...

1) According to this theory, all the matter and energy in the Universe must have been compressed into a very small space. Then it exploded from that single 'point' and started expanding.
2) The expansion is still going on. We can use the current rate of expansion of the Universe to estimate its age. Our best guess is that the Big Bang happened about 14 billion years ago.
3) The Big Bang isn't the only game in town. The 'Steady State' theory says that the Universe has always existed as it is now, and it always will do. It's based on the idea that the Universe appears pretty much the same everywhere. This theory explains the apparent expansion by suggesting that matter is being created in the spaces as the Universe expands. But there are some big problems with this theory.

Hubble bubble toil and trouble...

Hubble was well clever — he sorted out the Curtis–Shapley squabble and made up a really useful law. He's so clever we named a jazzy space telescope after him (see page 127). So don't worry if this all seems a bit confusing — just keep trying and soon you'll be master of the Universe. Well, master of Hubble's law anyway.

The Origin and Fate of the Universe

Once upon a time there was a really Big Bang — that's the most convincing theory we've got.

The CMBR provides Strong Evidence for the Big Bang Theory

1) Scientists have detected low-frequency electromagnetic radiation coming from all parts of the Universe.
2) This radiation is largely in the microwave part of the EM spectrum (p.32). It's known as the cosmic microwave background radiation (CMBR).
3) The Big Bang theory (see previous page) is the only theory that explains the CMBR.
4) Just after the Big Bang while the Universe was still extremely hot, everything in the Universe emitted very high-frequency radiation. As the Universe expanded it has cooled, and this radiation has dropped in frequency and is now seen as microwave radiation.

The Big Bang Theory Has Its Limitations

1) Today nearly all astronomers agree there was a Big Bang.
2) But the Big Bang theory isn't perfect. As it stands, it's not the whole explanation of the Universe — there are observations that the theory can't yet explain. E.g. the Big Bang theory predicts that the Universe's expansion should be slowing down — but as far as we can tell it's actually speeding up.
3) The Big Bang explains the Universe's expansion well, but it isn't an explanation for what actually caused the explosion in the first place, or what the conditions were like before the explosion (or if there was a 'before').
4) It seems likely the Big Bang theory will be adapted in some way to improve it, rather than just being dumped — it explains so much so well that scientists will need a lot of persuading to drop it altogether.

We Don't Know How (or If) the Universe Will End...

1) The Universe's ultimate fate depends on how fast it's expanding and the total mass there is in it. (The mass affects the gravitational pull that stops the Universe expanding so quickly.)
2) But these things are hard to measure, so determining the fate of the Universe is difficult.
3) To calculate how fast it's moving, you need to measure large distances, but the Universe is huge, so it's hard to accurately measure the distances involved.
4) You also need to accurately observe the motion of objects (e.g. galaxies). This is difficult because they're far away, you have to make lots of assumptions about their motion, and pollution gets in the way (p.128).
5) It's also tricky to measure how much mass there is because most of it appears to be invisible. Astronomers can only detect this dark matter by the way it affects the movement of the things we can see.

- If there's enough mass compared to how fast the galaxies are currently moving, the Universe will eventually stop expanding — and then begin contracting. This would end in a Big Crunch.
- If there's not enough mass in the Universe to stop the expansion, it could expand forever, with the Universe becoming more and more spread out into eternity.

Time and space — it's funny old stuff isn't it...

Proving a scientific theory is impossible. If enough evidence points a certain way, then a theory can look pretty convincing. But that doesn't prove it's a fact — new evidence may change people's minds.

Space Telescopes

The atmosphere might be good for us (blocking out X-rays for instance), but it's a pain for astronomers...

The Atmosphere Can Mess Up Measurements

1) Astronomers need accurate measurements to be able to understand what's going on in space, but our atmosphere can muck up the results.
2) Our atmosphere only lets certain wavelengths of electromagnetic radiation through and blocks all the others. The graph shows how the transparency of the atmosphere varies with wavelength.

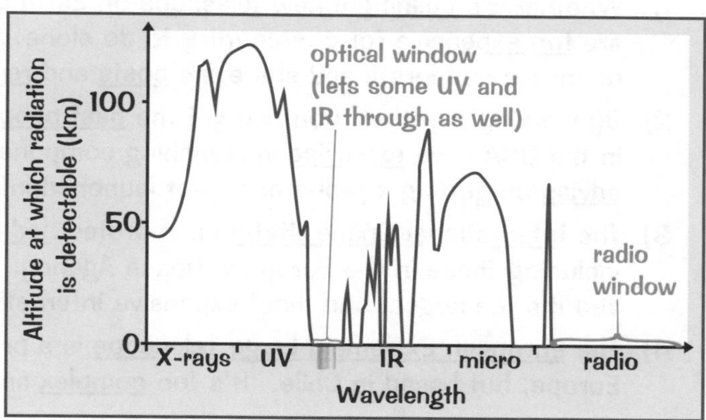

3) Some radiation, like radio waves, passes through the atmosphere without much trouble, but visible light can be badly affected.
4) Light gets refracted by water in the atmosphere, which blurs the images.
5) Light can also be absorbed by dust particles in the air. Boo hiss.
6) Sites for astronomical observatories on Earth are picked very carefully to try and minimise all these problems (see next page). Another solution is to take measurements from above the atmosphere...

Space Telescopes Have a Clearer View Than Those on Earth

1) If you're trying to look at EM radiation that's blocked or affected by the atmosphere, the thing to do is put your telescope in space, away from the mist and murk down here. The first space telescope (called Hubble) was launched by NASA in 1990. It can see objects that are about a billion times fainter than you can see just by standing in your back garden and looking up.

NASA/SCIENCE PHOTO LIBRARY

2) It's not all plain sailing though. Getting a telescope safely into space is hard. And when things go wrong, it's difficult to get the repair men out. Hubble's first pictures were all fuzzy, because the mirror was the wrong shape. NASA had to send some astronauts up there to fix it. D'oh.
3) Most astronomy is still done using Earth-based telescopes as they're a lot cheaper and easier to build and maintain. Hubble has cost over £3 billion to build, maintain and repair and people have to be sent to space to fix it.
4) Astronomers have also developed good techniques to remove the effects of the atmosphere from their measurements so the images are clearer.
5) It's also a lot easier to get a time slot to do your observing on Earth-based telescopes — there are lots more of them so there's much less demand for each one.

There are Many Uncertainties in Space Programmes

1) Space programmes are projects to send things like people, probes and telescopes into space. They're really expensive — the Apollo programme that ended with people walking on the Moon cost about $135 billion in today's money.

2) Governments have to balance paying these sums with other costly priorities like defence, healthcare and coping with natural disasters (which are unpredictable). The funding for space programmes is never guaranteed — there can be cut-backs at any time.
3) Many countries' space programmes are linked (see next page), so cut-backs in one country can have a knock-on effect on the others.

Telescope broken — we can't get the van up there, mate...

Space telescopes are so expensive because you've got to make a telescope that will be strong enough to withstand being blasted into space, but that's lightweight enough too. It's hard work being a boffin.

Observatories and Cooperation

It's just like primary school — astronomers have to learn to play together and share their toys.

Astronomers Need to Work Together

1) Whether it's building a new telescope on Earth or sending people into space, many science projects are too expensive for one country to do alone. These 'big science' projects are only possible if several countries cooperate and share the costs and resources.
2) By working together, you can get the best people and the best facilities for the job. E.g. scientists in the USA have expertise in launching components and equipment and so would probably be best to advise on getting a probe or object launched into space.
3) The International Space Station is a project led by the US but with the help of 15 other countries, including those in the European Space Agency. Each country is providing different parts of the Station and it's the largest and most expensive international science project in history.
4) The European Extremely Large Telescope is a project involving astronomers from across the whole of Europe, but based in Chile. It's too complex and expensive for a single country to build and operate.

Observatory Locations are Chosen for Astronomical Reasons...

1) Optical (visible light) observatories are often put in remote locations, e.g. Roque de los Muchachos in the Canary Islands. The idea is to avoid man-made light pollution (e.g. from street lamps) as well as dust and other particles (e.g. from car exhausts) affecting the observations.
2) Astronomers want as little atmosphere between the observatory and telescope as possible to minimise the distorting and blurring effects it has. So observatories are often built at high elevation (i.e. up mountains) where the atmosphere is thinner and so affects the light less. E.g. the Mauna Kea site in Hawaii is about 4200 m above sea level.
3) Water in the atmosphere can cause problems by refracting light — so a dry location with low atmospheric pollution is good for a telescope. E.g. there are observatories in Australia and the Atacama desert in Chile.
4) Clouds block a telescope's view of the sky, so they're built in places that have loads of cloudless nights.

MAGRATH PHOTOGRAPHY/ SCIENCE PHOTO LIBRARY

... But Other Factors Need to be Taken into Account

Scientists have to live in the real world. They can't just go building observatories in the most remote and awkward places, just because there's a nice view. There are other things to take into account...

1) COST — Observatories aren't cheap to build. Never mind carting all the stuff up a mountain in the middle of nowhere. There's the cost of building, running and eventually closing the observatory.

2) ACCESS — The site will have to have roads built to it (so you can get equipment and people there to build the telescopes) as well as electricity and other facilities. Some places are just too hard to get to.

3) ENVIRONMENT — Scientists have to be careful that building works, etc. will damage the surrounding environment as little as possible, e.g. by disturbance to wildlife or agriculture.

4) SOCIAL — Even with remotely controlled telescopes, there are always going to be a few people who need to work at the telescope site. They'll need facilities such as water, electricity, accommodation, shops, etc., which will be quite expensive to provide. In some areas observatories have benefited the local community — by providing jobs in building and maintaining the observatory.

Hawaii? — Sounds like a sneaky holiday to me...

Most of the big optical and infrared observatories on Earth are in Hawaii, Chile, Australia and the Canary Islands.

Revision Summary for Section Ten

There's a lot of tricky stuff to learn in this section, but at least there are lots of pretty pictures to help. Have a go at these questions, and if you get stuck have a sneaky peek back at the pages. Just keep going through them till you can do them all.

1) Draw a labelled diagram of the Earth showing the crust, mantle and core.
2) What causes tectonic plates to drift?
3) What causes earthquakes?
4) Briefly explain Wegener's theory of continental drift.
5) Briefly explain the changes you see in the magnetic orientation of sea floor rocks.
6) Name the two types of seismic waves caused by earthquakes, and state whether each type is a transverse or longitudinal wave.
7) What type of seismic wave cannot pass through liquids?
8) Describe the Copernican model of the Universe.
9) Explain the evidence that Galileo produced that supported Copernicus' theory.
10) How many planets are in the Solar System?
11) Other than planets, name two things that orbit the Sun.
12) What are comets made of?
13) Roughly how many stars make up the Milky Way?
14) What is a light year?
15) Roughly how many times larger is the diameter of the Milky Way than the diameter of the Sun?
16) List the steps that lead to the formation of a main sequence star (like our Sun).
17) A star much larger than the Sun stops fusing hydrogen. Describe the phases the star will go through before becoming a neutron star or a black hole.
18) Give the two ways in which energy is transported to the surface of a star.
19) What is the 'peak frequency' of a star?
20) Explain how absorption line spectra can show what elements are present in a star's atmosphere.
21) Briefly explain what a sidereal day is.
22) State what a solar day is.
23) Describe how a lunar eclipse occurs.
24) Name the two angles which are used to measure the positions of stars in the sky.
25) State what the parallax angle is.
26) Briefly outline Curtis' argument in the Curtis-Shapley debate.
27) If a wave source is moving towards you, will the observed frequency of its waves be higher or lower than their actual frequency?
28) What do red-shift observations tell us about the Universe?
29)*The Tadpole galaxy is approximately 4×10^{21} km away from Earth. Calculate its speed of recession, if the Hubble constant is 2×10^{-18} s^{-1}.
30) Describe the 'Big Bang' theory for the origin of the Universe. What evidence is there for this theory?
31) Explain why astronomers can only detect certain wavelengths of radiation from space on Earth.
32) Give one advantage of using space telescopes.
Give two reasons why most astronomy is done with Earth-based telescopes.
33) Why do astronomers need to work together on 'big science projects'?
34) Write down two astronomical and two non-astronomical factors that have to be taken into account when choosing a site for an observatory.

* Answers on page 140.

Potential Dividers

Potential dividers consist of a pair of resistors. They divide the potential in a circuit so you can get outputs of different voltages.

The Higher the Resistance, the Greater the Voltage Drop

A voltage across a pair of resistors is 'shared out' according to their relative resistances. The rule is:

The larger the share of the total resistance, the larger the share of the total voltage.

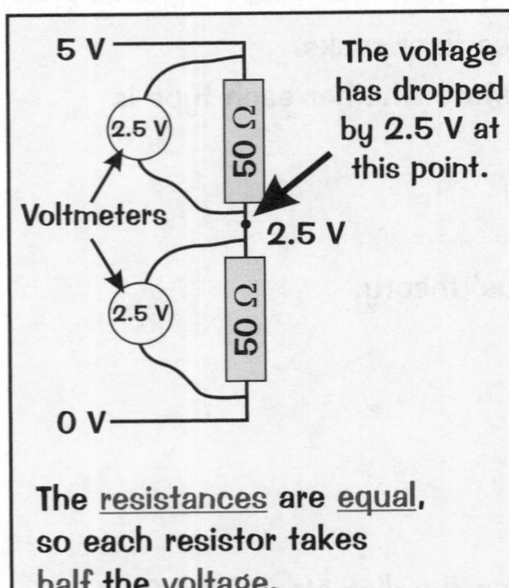

The resistances are equal, so each resistor takes half the voltage.

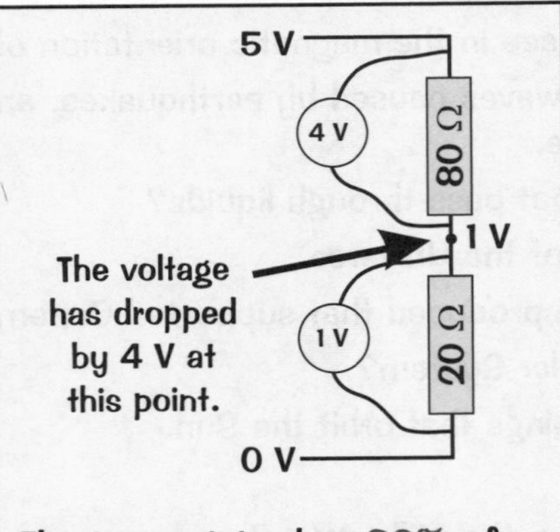

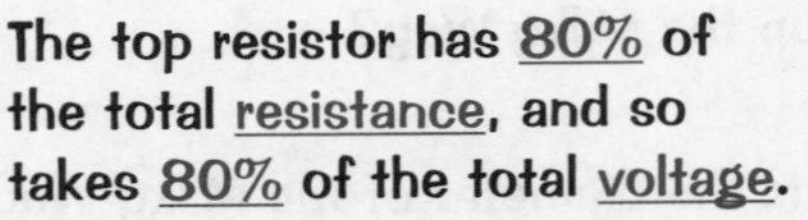

The top resistor has 80% of the total resistance, and so takes 80% of the total voltage.

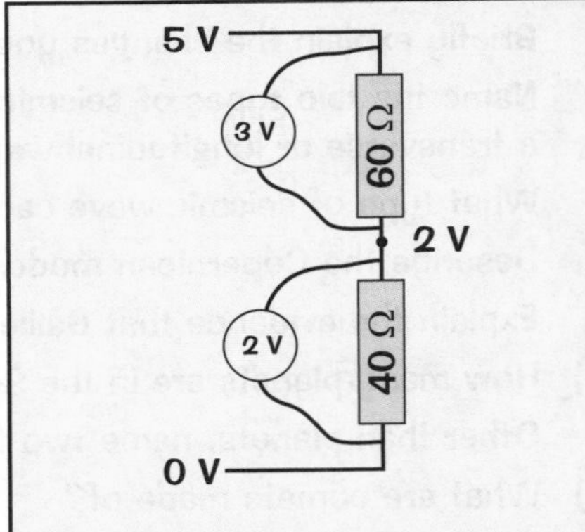

The top resistor has 60% of the total resistance, and so takes 60% of the total voltage.

The point between the two resistors is the 'output' of the potential divider. This 'output' voltage can be varied by swapping one or both of the resistors for a variable resistor.

Potential Dividers are Quite Useful

Potential dividers are not only spectacularly interesting — they're useful as well. They allow you to run a device that requires a certain voltage from a battery of a different voltage. This is the formula you need to use:

$$V_{out} = V_{in} \times \left(\frac{R_2}{R_1 + R_2}\right)$$

The output voltage (V_{out}) depends on the relative values of R_1 and R_2. From the formula, you should see that if R_2 is very big compared to R_1, the bit in the brackets cancels down to about 1, so V_{out} is approximately V_{in}. But if R_2 is a lot smaller than R_1, the bit in brackets becomes so small that V_{out} is approximately 0.

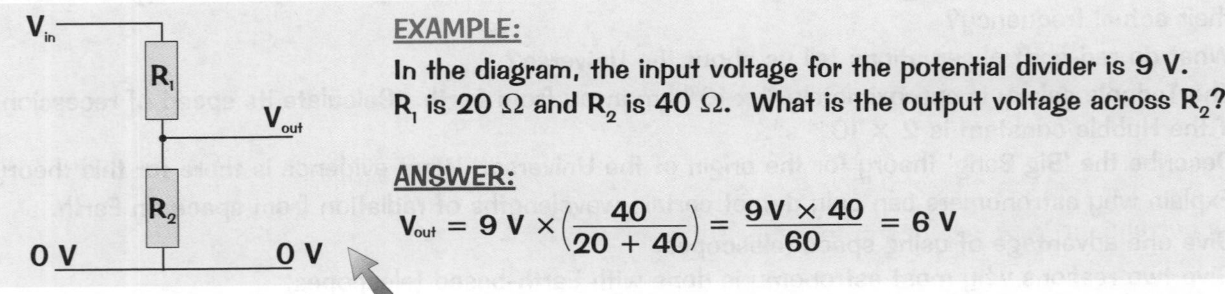

EXAMPLE:

In the diagram, the input voltage for the potential divider is 9 V. R_1 is 20 Ω and R_2 is 40 Ω. What is the output voltage across R_2?

ANSWER:

$$V_{out} = 9\text{ V} \times \left(\frac{40}{20 + 40}\right) = \frac{9\text{ V} \times 40}{60} = 6\text{ V}$$

A potential divider like this could be used to run a 6 V device from a 9 V battery. You could replace one of the resistors by a variable resistor, so that you could change V_{out} to any value between 0 and 9 volts. If you had two variable resistors you could have much finer control of an output voltage with an adjustable threshold — useful for controlling heat and light sensors (see next page).

My boyfriend's mother is a potential divider...

You won't believe this, but potential dividers get even more exciting on the next page. I know you're worried that you won't be able to cope with the adrenaline rush but it's just something you have to get used to with Physics...

Thermistors and Transistors

A thermistor is a resistor that changes its resistance depending on the temperature (see p.60).

A Thermistor in a Potential Divider Makes a Temperature Sensor

1) Using a thermistor and a fixed resistor in a potential divider, you can make a temperature sensor.
2) You can make a temperature sensor that gives a high voltage output (a 'logical 1' — see page 132) when it's hot and a low voltage output (a 'logical 0') when it's cold. This is how it works...

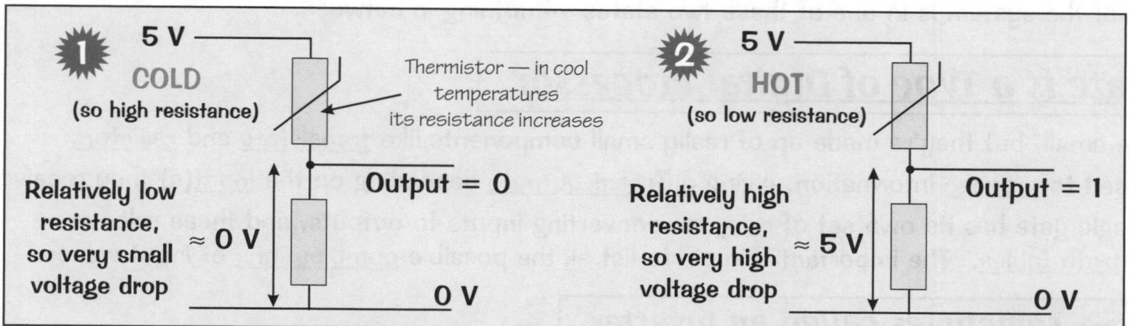

1 When the thermistor's cold its resistance is very high, so the voltage drop across it is almost 5 V, meaning the voltage of the output is nearly 0 V — a 'logical 0'.

2 As the temperature of the thermistor increases, its resistance falls dramatically. So the voltage across it is almost 0 V and the voltage of the output is nearly 5 V — a 'logical 1'.

3) In the circuit above, the thermistor is in the R_1 position (see previous page). If you switched the circuit around so the thermistor was in the R_2 position and the resistor was in the R_1 position then the output would switch round too. So, the output would be 1 when it's cold and 0 when it's hot.
4) If you replace the fixed resistor with a variable resistor, you can make a sensor that triggers an output device at a temperature you choose and can change whenever you like, e.g. in a heating system.

Transistors are Electronic Switches

Transistors are the basic building blocks of electronic components — modern computers contain billions of them.

1) In transistors, a small amount of current is used to control the flow of a much larger current. This means that they can be used as electronic switches.
2) Transistors can be much smaller than mechanical switches, so they can be integrated into circuits, such as logic gates.
3) All transistors have three parts:
   a) Base — The 'switch' that controls the flow of current. If no current is applied to the base it stops current flowing through the rest of the transistor. When a small current is applied to the base a larger current can flow through the collector and emitter. A large current passing through the base would damage the transistor.
   b) Collector — Current flows into the transistor through the collector.
   c) Emitter — Current flows out of the transistor through the emitter.

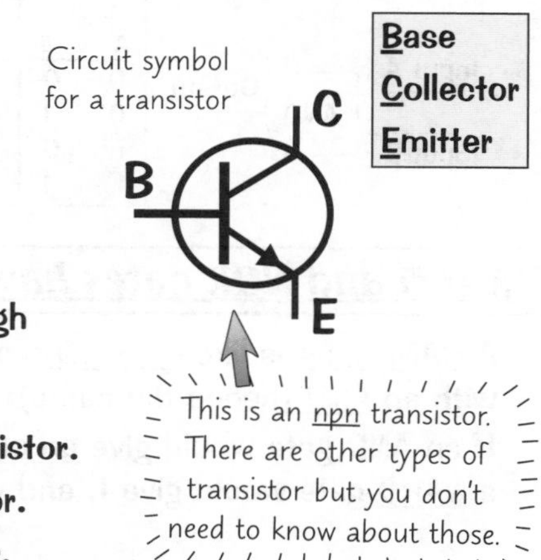

4) The currents in each part of the transistor are related by this handy equation:

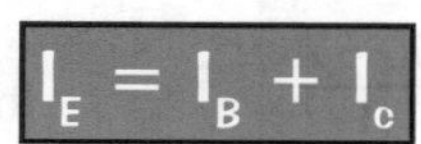

$$I_E = I_B + I_C$$

Current in emitter = current in base + current in collector

EXAMPLE: A current of 0.1 A is applied to the base of a transistor. This allows a current of 2 A to flow through the collector. Calculate the current which flows through the emitter.

ANSWER: Use the equation $I_E = I_B + I_C$. $I_E = 0.1\text{ A} + 2\text{ A} = 2.1\text{ A}$.

From a tenement window a transistor [radio] blasts...

Transistors can also be used to amplify a signal, so the first common devices to use transistors were radios. Before transistors, radios used bulky and fragile vacuum tubes. Transistors made radios portable and cheap.

Logic Gates

Transistors can also be combined to make logic gates, which process information and make computers work.

Digital Systems are Either On or Off

1) Every connection in a digital system is in one of only two states. It can be either ON or OFF, either HIGH or LOW, either YES or NO, either 1 or 0... you get the picture.
2) In reality a 1 is a high voltage (about 5 V) and a 0 is a low voltage (about 0 V). Every part of the system is in one of these two states — nothing in between.

A Logic Gate is a Type of Digital Processor

Logic gates are small, but they're made up of really small components like transistors and resistors.

They can be used to process information, giving different outputs depending on the input(s) they receive.

Each type of logic gate has its own set of rules for converting inputs to outputs, and these rules are best shown in truth tables. The important thing is to list all the possible combinations of input values.

NOT gate — sometimes called an Inverter

A NOT gate just has one input — and this input can be either 1 or 0, so the truth table has just two rows.

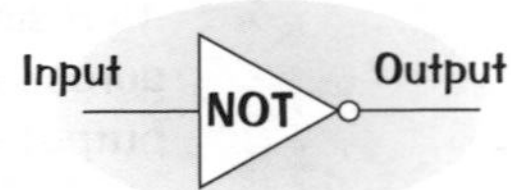

NOT GATE

Input	Output
0	1
1	0

Some AND and OR gates have more than two inputs, but you don't have to worry about those.

AND and OR gates usually have Two Inputs

Each input can be 0 or 1, so to allow for all combinations from two inputs, your truth table needs 4 rows. There's a certain logic to the names...

An AND gate only gives an output of 1 if both the first input AND the second input are 1.

Input A
AND
Output
Input B

AND GATE

Input A	Input B	Output
0	0	0
0	1	0
1	0	0
1	1	1

An OR gate just needs either the first OR the second input to be 1.

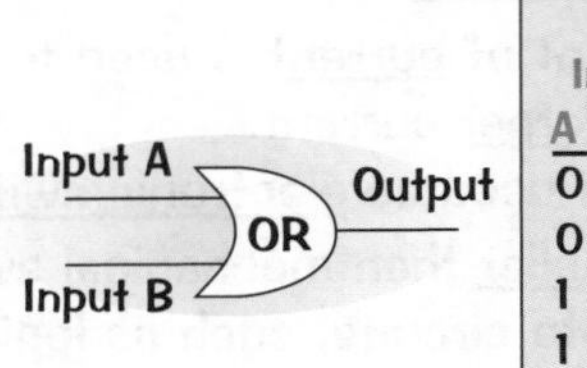

OR GATE

Input A	Input B	Output
0	0	0
0	1	1
1	0	1
1	1	1

NAND and NOR gates have the Opposite Output of AND and OR gates

A NAND gate is like combining a NOT with an AND (hence the name): If an AND gate would give an output of 0, a NAND gate would give 1, and vice versa.

Input A
NAND
Output
Input B

NAND GATE

Input A	Input B	Output
0	0	1
0	1	1
1	0	1
1	1	0

A NOR gate is like combining a NOT with an OR (hence the name): If an OR gate would give an output of 0, a NOR gate would give 1, and vice versa.

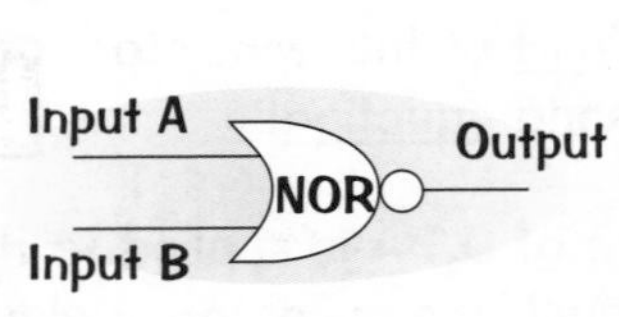

NOR GATE

Input A	Input B	Output
0	0	1
0	1	0
1	0	0
1	1	0

I like physics, NAND chemistry, NAND biology...

Well at least there aren't that many facts to learn on this page — it's more a question of understanding the inputs and outputs for the five types of gate. It's a good idea to be familiar with the circuit symbols of the gates too though. And practise writing out all the different tables — it's the quickest and bestest way to learn.

Using Logic Gates

You might need to construct a truth table for a combination of logic gates.
Approach this kind of thing in an organised way and stick to the rules, and you won't go far wrong.

'Interesting' Example — a Greenhouse

Once the gardener has switched the system on, he wants to be warned if the greenhouse gets too cold or if someone has opened the door. He only wants the warning system to work when the greenhouse gets dark.

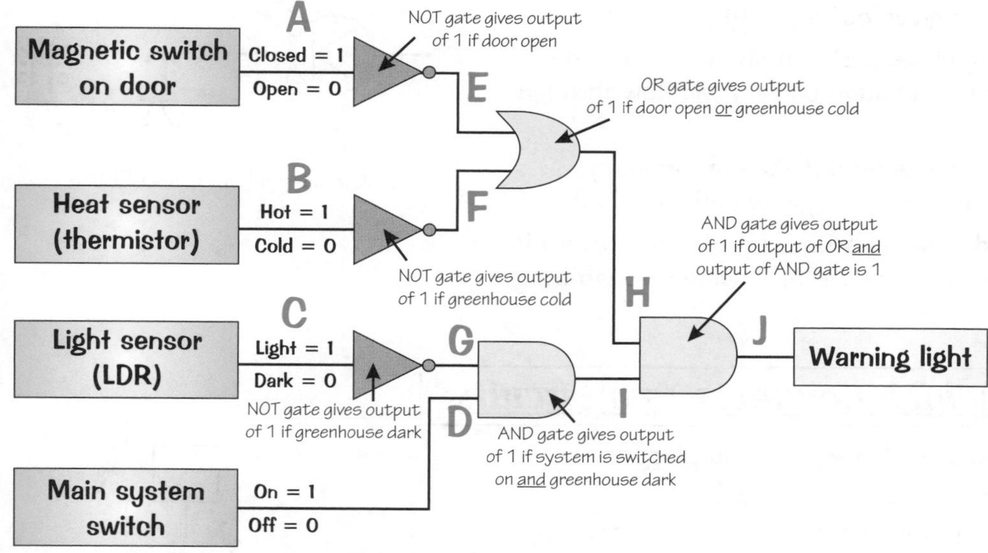

Inputs									Output
A	B	C	D	E	F	G	H	I	J
0	0	0	0	1	1	1	1	0	0
0	0	0	1	1	1	1	1	1	1
0	0	1	0	1	1	0	1	0	0
0	0	1	1	1	1	0	1	0	0
0	1	0	0	1	0	1	1	0	0
0	1	0	1	1	0	1	1	1	1
0	1	1	0	1	0	0	1	0	0
0	1	1	1	1	0	0	1	0	0
1	0	0	0	0	1	1	1	0	0
1	0	0	1	0	1	1	1	1	1
1	0	1	0	0	1	0	1	0	0
1	0	1	1	0	1	0	1	0	0
1	1	0	0	0	0	1	0	0	0
1	1	0	1	0	0	1	0	1	0
1	1	1	0	0	0	0	0	0	0
1	1	1	1	0	0	0	0	0	0

The warning light will come on if:

i) it is cold in the greenhouse OR if the door is opened,
ii) AND the system is switched on,
iii) AND the greenhouse is dark.

1) Each connection has a label, and all possible combinations of the inputs are included in the table.
2) What really matters are the inputs and the output — the rest of the truth table is just there to help.
3) An LDR (see p.60) or thermistor combined with a resistor makes a light or temperature sensor (see page 131), the output of which can produce an input signal for a logic circuit (as used above).
4) The resistance changes the 'threshold voltage' (i.e. how bright or hot it needs to be to produce a signal).
5) Using a variable resistor makes the threshold voltage adjustable — e.g. the gardener can adjust the temperature the warning light comes on at.

AND Logic Gates are Made From Two Transistors

1) AND logic gates give an output of 1 if both inputs are also 1.
2) AND gates are made using a series of two transistors.
3) Each input is connected to the base of a transistor.
4) If the signal of either input is 0, no current flows through the base of the transistor it's connected to. The transistor stays open, so no current can flow through the rest of the circuit and the output of the gate will be 0.
5) If both inputs are 1 then both transistors will be closed, current will flow through the gate and the output of the gate will be 1.
6) Other logic gates can be made from different combinations of two transistors.

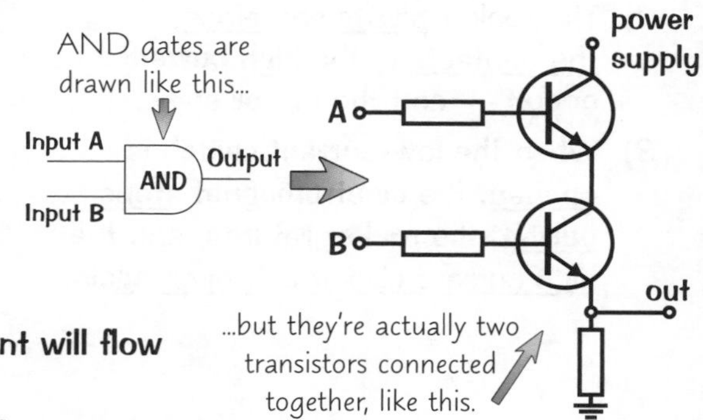

Now we can all sleep easy knowing the cucumbers are safe...

More hard stuff to get your head around here. Try copying out the diagrams of the greenhouse warning system and writing in the different inputs. Then follow them through to find the outputs. It's almost fun. Almost.

LEDs and Relays in Logic Circuits

Two main points on this page: 1) An LED can be used to display the output of a logic gate. 2) Logic gates don't usually supply much current, so they're often connected to a more powerful circuit using a relay switch.

LEDs — Light-Emitting Diodes

1) An LED is a diode (see page 135) which gives out light.
2) Like other diodes, it only lets current go through in one direction. When it does pass current, it gives out a pretty coloured light.
3) You can use a light-emitting diode (LED) to show the output of a logic gate. If the output is 1, enough current will flow through the LED to light it up.
4) An LED is a better choice to show output than an ordinary incandescent bulb because it uses less power and lasts longer.
5) The LED is often connected in series with a resistor to prevent it from being damaged by too large a current flowing through it.

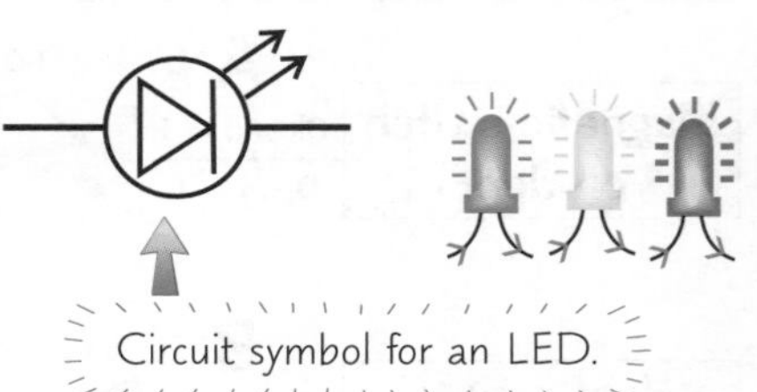

Circuit symbol for an LED.

A Relay is a Switch Which Connects Two Circuits

1) A low-power logic gate would be damaged if you plugged it straight into a high current mains power supply.
2) But an output device like a motor requires a large current.
3) The solution is to have two circuits connected by a relay.
4) The relay isolates the low voltage electronic system from the high voltage mains often needed for the output device.
5) This also means that it can be made safer for the person using the device — you can make sure that any parts that could come into contact with a person are in the low-current sensing circuit. For example, a car's starter motor needs a very high current, but the part you control (when you're turning the key) is in the low-current circuit — safely isolated by the relay.

There are a few circuit symbols for a relay — this is the simplest one.

Here's How a Relay Works...

1) When the switch in the low-current circuit is closed, it turns on the electromagnet (see page 81), which attracts the iron contact on the rocker.
2) The rocker pivots and closes the contacts in the high current circuit — and the motor spins.
3) When the low-current switch is opened, the electromagnet stops pulling, the rocker returns, and the high current circuit is broken again.

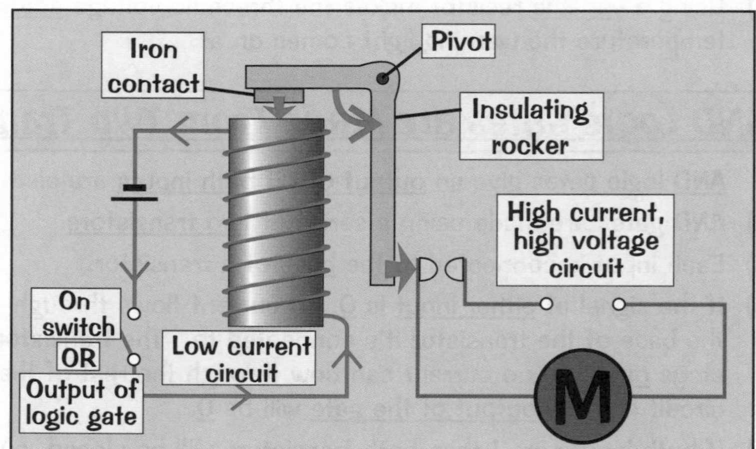

You should now be relay proud of yourself...

...'cos you've managed to get through all those pages on logic gates. I know it isn't always a barrel of laughs. There's a lot of tricky stuff here — a lot to learn, and a lot that's hard to understand. It's definitely a good idea to learn each page thoroughly before moving on, otherwise it'll all turn into a big tangled mess in your brain.

Diodes and Rectification

Mains electricity supplies alternating current (AC), but many devices need direct current (DC). So we need a way of turning AC into DC. That's where diodes come in.

Diodes Only Let Current Flow in One Direction

1) Diodes only let current flow freely in one direction — there's a very high resistance in the other direction.
2) You can tell which direction the current can flow through a diode from the circuit symbol.

   The triangle points in the direction of the current.

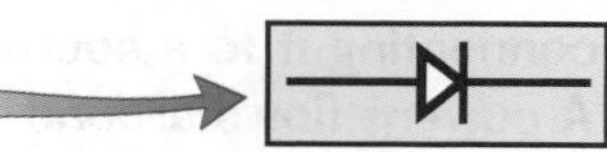

Here the current flows from left to right.

Diodes are Made from Semiconductors Such As Silicon

1) Diodes are often made of silicon, which is a semiconductor. This means silicon can conduct electricity, though not as well as a conductor.
2) Silicon diodes are made from two different types of silicon joined together at a 'p-n junction'. One half of the diode is made from silicon that has an impurity added to provide extra free electrons — called an n-type semiconductor ("n" stands for the "negative" charge of the electrons).
3) A different impurity is added to the other half of the diode so there are fewer free electrons than normal. There are lots of empty spaces left by these missing electrons which are called holes. This type of silicon is called a p-type semiconductor ("p" stands for the "positive" charge of the holes).
4) When there's no voltage across the diode, electrons and holes recombine where the two parts of the diode join. This creates a region where there are no holes or free electrons, which acts as an electrical insulator.
5) When there is a voltage across the diode the direction is all-important: Applying a voltage in the RIGHT direction means the free holes and electrons have enough energy to get across the insulating region to the other side. This means that a CURRENT FLOWS. Applying a voltage in the WRONG direction means the free holes and electrons are being pulled away from the insulating region, so they stay on the same side and NO CURRENT FLOWS.

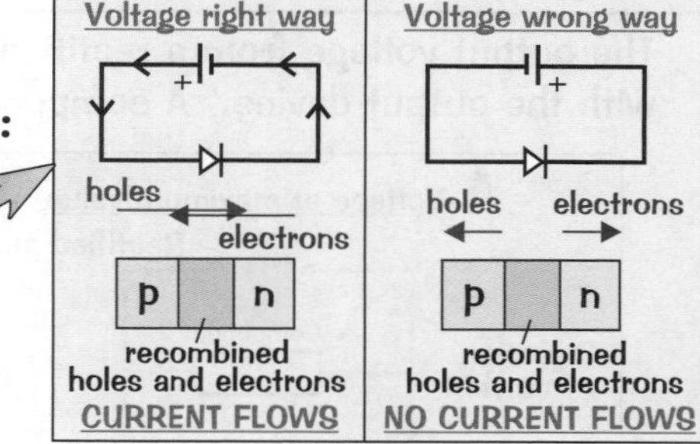

Diodes Can be Used to Rectify Alternating Current

1) A single diode only lets through current in half of the cycle. This is called half-wave rectification.

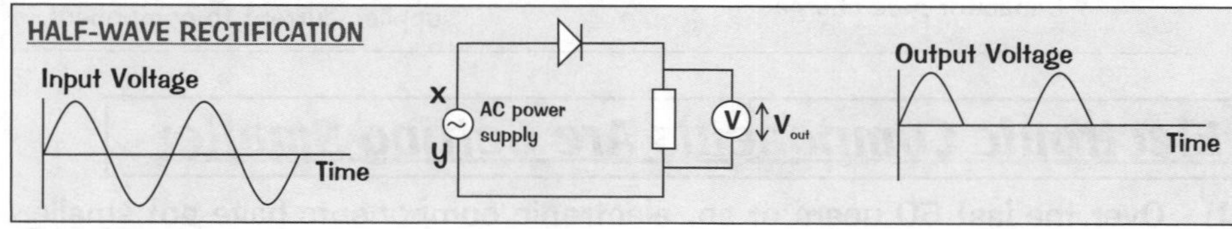

2) To get full-wave rectification, you need a bridge circuit with four diodes. In a bridge circuit, the current always flows through the component in the same direction, and the output voltage always has the same sign.

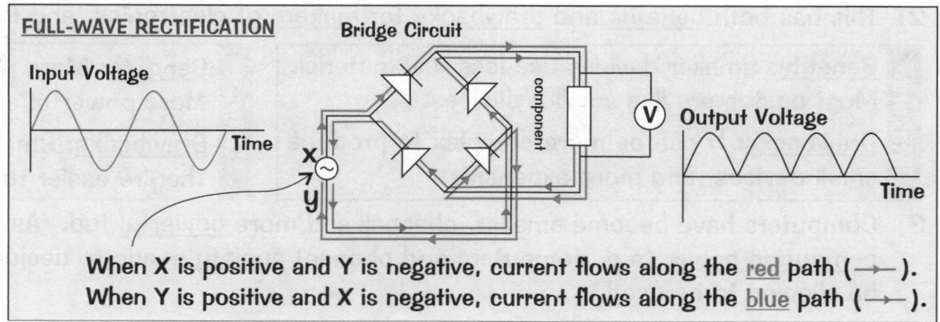

When X is positive and Y is negative, current flows along the red path (—→—).
When Y is positive and X is negative, current flows along the blue path (—→—).

Yep, it's all just common sense really...

Only joking — this stuff's flippin' hard. At least you've made it through to the other side though, well done.

Capacitors

AC voltage that has been rectified is not all that useful in its raw form. For example, computer chips are very sensitive to input voltage, and won't work with a voltage that looks like this: .
They need a smoother voltage like this: . This is where capacitors come in handy.

Capacitors Store Charge

1) You charge a capacitor by connecting it to a source of voltage, e.g. a battery. A current flows around the circuit, and charge gets stored on the capacitor.
2) The flow of current decreases as you charge for longer periods of time.
3) The more charge that's stored on a capacitor, the larger the voltage across it.
4) When the voltage across the capacitor is equal to that of the battery, the current stops and the capacitor is fully charged. The voltage across the capacitor won't rise above the voltage of the battery.
5) If the battery is removed, the capacitor discharges — the flow of current is the same for discharging as for charging (see shape of graph above) but the current flows in the opposite direction round the circuit.

Circuit symbol

Current in A

Capacitor

Time

CHARGING:

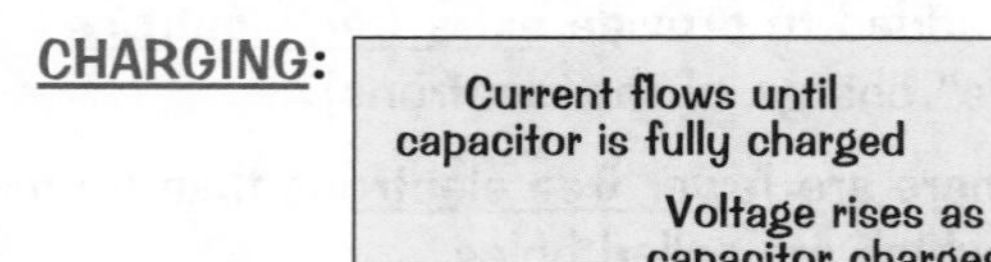

DISCHARGING:

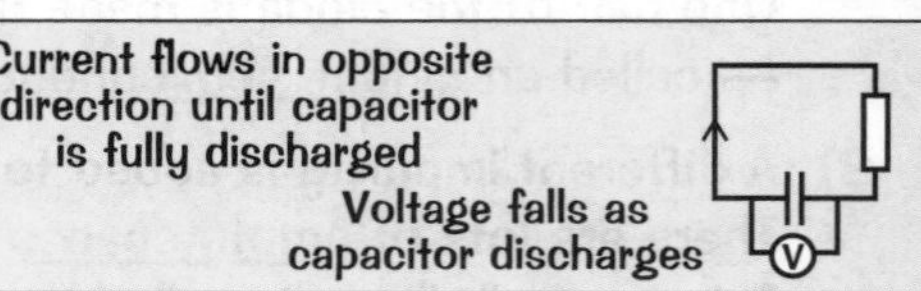

Capacitors are Used in 'Smoothing' Circuits

The output voltage from a rectified AC power supply can be 'smoothed' by adding a capacitor in parallel with the output device. A component gets current alternately from the power supply and the capacitor.

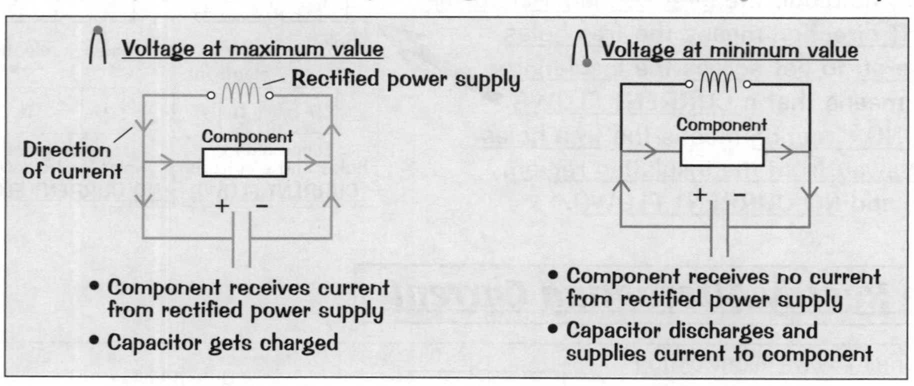

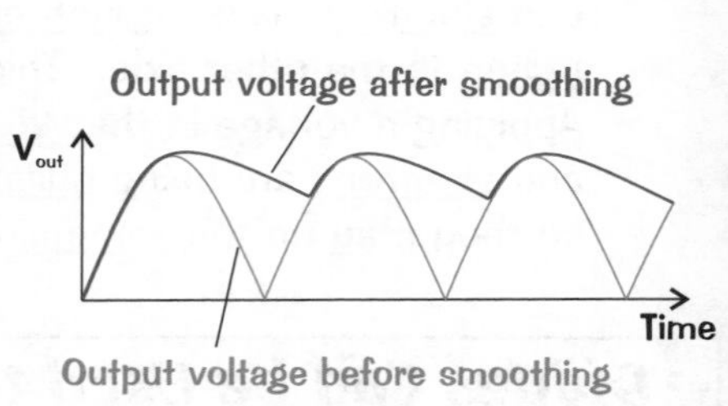

Electronic Components Are Getting Smaller

1) Over the last 50 years or so, electronic components have got smaller — it's known as miniaturisation.
2) This has both benefits and drawbacks for makers of electronics, and for the people using them:

Makers	Benefits: Smaller devices use less raw material. Most customers like smaller devices. Drawbacks: It can be more complex to produce small devices, and more expensive.
Users	Benefits: More portable electronic devices available. More powerful and feature-filled devices produced. Drawbacks: Smaller devices can be more expensive. They're easier to lose down the back of the sofa.

3) Computers have become smaller, cheaper and more powerful too. As more people have access to more computing power (e.g. computers and phones) society needs to decide how this should, or should not, be allowed to be used.

   For example, society needs to think about controlling hacking, piracy and access to personal data.

Current never flows through a capacitor...

Capacitors just store charge, and then send current back the other way when the voltage falls. Handy.

Revision Summary for Section Eleven

Oooo — a whole section of actually useful, real-worldy stuff. Gosh, it's a bit tricky though, isn't it... Do these questions to see how you're getting on. Then you can relax, it's the end of the book. Ace.

1) Explain how potential dividers work.
2) State the formula for potential dividers.
   Do your own worked example, including a sketch.
3)* The diagram shows a potential divider with an input voltage of 9 V. R_1 is 10 Ω and R_2 is a variable resistor. Calculate the output voltage across R_2 when a) R_2 = 30 Ω b) R_2 = 2 Ω.
4) Explain how you would use a thermistor in a potential divider to make a temperature sensor.
5) Draw and label the circuit symbol for a transistor.
6) Describe what happens in each part of a transistor.
7)* A current of 0.3 A is applied to the base of a transistor, allowing a current of 5 A to flow through the collector. Calculate the size of the current that flows through the emitter.
8) Draw truth tables for AND, OR, NAND and NOR gates.
9)* The diagram below shows how logic gates can be used to monitor the temperature inside a greenhouse.

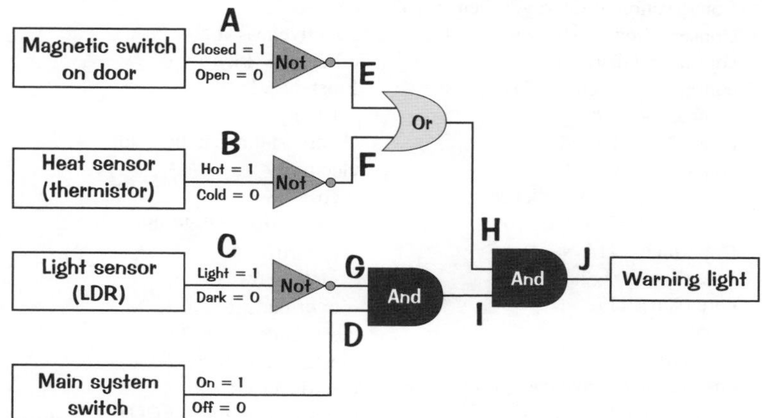

   The warning light will come on if the output at J is 1.
   For the following input, work out whether the warning light is on or off.
   A = 1, B = 0, C = 1, D = 1
10) What are AND gates made from?
11) Explain how an LED can be used to show the output of a logic gate.
12) Make a sketch of a relay.
13) Explain briefly how a diode works.
14) What semiconducting material are diodes often made of?
15) Explain the two ways in which an AC current can be rectified.
    Include circuit diagrams and voltage/time graphs in your explanation.
16) What is a capacitor?
17) How can a capacitor be used to smooth rectified voltage?
18) Electronic components are getting smaller. Explain why this could be a bad thing for:
    a) manufacturers of electronic devices,
    b) users of electronic devices.
19) Go and read up on quantum theory... no wait, I mean... go and put the kettle on.

* Answers on page 140.

Index

Index

Index

Answers

Revision Summary for Section One (page 26)

14) 40 years
19) 90 kJ
25) 70% or 0.7
26) a) 80 J b) 20 J c) 0.8 or 80%

Wave Basics Top Tip (page 27)

2375 m/s

Revision Summary for Section Two (page 39)

3) 150 m/s

Revision Summary for Section Three (page 54)

3) a = (v – u) ÷ t; 35 m/s^2
11) F = ma, so a = F ÷ m; 7.5 m/s^2
12) 1.33 m/s^2
13) 120 N
18) E = F × d; 6420 J
19) KE = ½ × m × v^2; 20 631 J
20) The car would stop in 5.1 m, so he will hit the sheep.
21) a) 1200 J
b) 600 J
22) 15 600 J

The Cost of Electricity Top Tip (page 68)

1) 2.76 kW
2) 11 minutes (to the nearest minute)

Revision Summary for Section Four (page 70)

6) 4 A
12) 1.2 V
13) 4.8 Ω
16) a) 1.2 A
b) 7.2 V
c) 12 V
19) 12.5 Hz
25) Hair straighteners: E = 13 500 J
Hair dryer: E = 12 600 J
The hair straighteners use more energy.
26) 3180 J
27) 0.125 kWh

Revision Summary for Section Five (page 80)

8) a) $^{131}_{53}\text{I} \rightarrow {}^{131}_{54}\text{Xe} + {}^{0}_{-1}\beta$

b) $^{148}_{64}\text{Gd} \rightarrow {}^{144}_{62}\text{Sm} + {}^{4}_{2}\alpha$

15) 1 hr 20 minutes

Revision Summary for Section Six (page 87)

14) 3.5 A
16) 270 V
17) 80 turns
21) 1408 W (1.41 kW)

Revision Summary for Section Seven (page 93)

9) 2 cm
23) 86 beats per minute

Revision Summary for Section Eight (page 101)

2) 770 Nm
4) 2 m
8) 0.1 s
11) 80 Pa
12) 2560 N
15) a) 184 K
b) 393 K
c) –268 °C
d) 39 °C
17) 405 300 Pa
19) 2.9 atm

Revision Summary for Section Nine (page 111)

1) 2.4
2) 1.47 (to 2 d.p.)
6) 41.1°
12) 3
13) 4 D
16) v = 0.6 m from the lens
23) 40

Revision Summary for Section Ten (page 129)

29) 8000 km/s

Revision Summary for Section Eleven (page 137)

3) a) 6.75 V
b) 1.5 V
7) 5.3 A
9) Output = 0, so the warning light is off.